PHILOPONUS

Against Proclus

On the Eternity of

the World 1-5

PHILOPONUS

Against Proclus On the Eternity of the World 1-5

Translated by
Michael Share

Duckworth

First published in 2004 by
Gerald Duckworth & Co. Ltd.
90-93 Cowcross Street, London EC1M 6BF
Tel: 020 7490 7300
Fax: 020 7490 0080
inquiries@duckworth-publishers.co.uk
www.ducknet.co.uk

A catalogue record for this book is available from the British Library

ISBN 0 7156 3233 7

Acknowledgements

The present translations have been made possible by generous and imaginative funding from the following sources: the National Endowment for the Humanities, Division of Research Programs, an independent federal agency of the USA; the Leverhulme Trust; the British Academy; the Jowett Copyright Trustees; the Royal Society (UK); Centro Internazionale A. Beltrame di Storia dello Spazio e del Tempo (Padua); Mario Mignucci; Liverpool University; the Leventis Foundation; the Arts and Humanities Research Board of the British Academy; the Esmée Fairbairn Charitable Trust; the Henry Brown Trust; Mr and Mrs N. Egon; the Netherlands Organisation for Scientific Research (NWO/GW); the Cultural Attaché of the Greek Embassy in London. The editor wishes to thank Frans de Haas, James Wilberding, and Andrea Falcon for their comments and John Bowin, Devin Henry and Inna Kupreeva for preparing the volume for press.

A special grant for the preparation of this volume for press was provided by the Council of Gresham College, to which the series editor is very grateful.

Typeset by Ray Davies
Printed and bound in Great Britain by
Biddles Ltd, King's Lynn, Norfolk

Contents

Preface

Richard Sorabji

This work in eighteen chapters, of which the first five are translated here, is one of the most interesting of all post-Aristotelian Greek philosophical texts, written at a crucial moment in the defeat of paganism by Christianity. In 529 AD, the Emperor Justinian put an end to teaching in the pagan Neoplatonist school in Athens, where Proclus had in the fifth century AD been the most devout pagan teacher, St Benedict is thought to have founded the monastery in Monte Cassino, and, again on behalf of Christianity, Philoponus in Alexandria attacked Proclus' arguments that the universe had no beginning in his *Against Proclus on the Eternity of the World*. Philoponus was one of the cleverest of the Neoplatonist philosophers, a pupil of Ammonius in Alexandria, but he was a Christian, and he used his profound knowledge of the Neoplatonist and Aristotelian traditions to turn the pagans' own views against themselves.

Our text records, and replies to, the 18 arguments of Proclus' *Against the Christians on the Eternity of the World*, as well as quoting a little of Proclus' *Examination of Aristotle's Objections to Plato's Timaeus*. It will suffice to indicate just a few of the original arguments and ideas in these chapters.

In I.3, pp. 9,14-11,17, Philoponus exploits Aristotle's concept of infinity, which the Neoplatonists accepted, as a weapon with which to refute them. His arguments, which were deployed in a number of his works, had already appeared in his commentary on Aristotle's *Physics*, originally written in 517 AD, whether or not subsequently revised, but there they were not very prominent. Aristotle in *Physics* 3.4-8, had allowed infinity to exist, but only in the way that it is still introduced in modern schools, as an ever expandable finitude. This meant that a more than finite number is never reached, which makes infinity seem more tractable. Aristotle put it by saying that infinities are potential, never actual, and that you can never go right through them. What Philoponus points out is that Christianity must, on this view, be right to say that the universe had a beginning, because otherwise it would have gone right through an actual infinity, a more than finite number, of years. Worse, it would have done so by the time of Socrates, who was executed earlier, in the fourth century BC, so since then it would have gone through more than an actual infinity. And if there had been an actual infinity of generations of humans, when you added in the generations of horses and dogs, then you too would have more than an actual infinity.

Aristotle had ruled out actual infinities in *Physics* 3.5, 204a20-6. There is an infinity of whole numbers in the sense that however large a finite number you have counted, more still can always be counted. But if you allow a more than finite number of whole numbers to exist, the even numbers will be as numerous as the odd and even combined, and part cannot be as large as whole. It was only in the fourteenth century that Western philosophers were able to explain the acceptable sense in which the set of whole numbers would be larger, and the sense in which it would not. I have described this elsewhere.[1] But in the first half of the thirteenth century, I am told, Grosseteste's treatise *On Light* seems to be aware of the viability of this, in which case I conjecture that he will have learnt it from Latin translations of Arabic sources, possibly of al-Haytham. At least, it does not seem to have been known to the Greeks, unless to Archimedes in the third century BC, who may have referred to the point in *The Method of Mechanical Theorems*, of which a fragment survives in a palimpsest that is currently being deciphered by Reviel Netz.

In IV.9, at pp. 78-9, Philoponus offers us another impressive argument. Although God only has to will in order to create the universe, and although he always possesses the blueprint for creating it, it does not follow that he should make it exist always. It might be less good that it should exist at all times, like himself. Philoponus here offers a version of the idea that Augustine had earlier expressed in Latin, that God's willing a change (the universe starting to exist) does not prevent his will being changeless. But Philoponus' inspiration will not have come from a Latin text. On p. 568 of *Against Proclus*, he is aware that his point about *willing* is analogous to the point made about *knowledge* by Proclus' hero, the third- to fourth-century pagan Neoplatonist Iamblichus, who said that knowledge does not have to have the same status as the known. Everlasting beings could have everlasting knowledge of the temporary.[2]

Proclus' 5th argument was that there was no time when there was not time, and the celestial clock of revolving stars, according to Plato's *Timaeus*, is co-extensive with time. Hence time and the heavens are both beginningless. Philoponus replies in V.4, 116,1-24, that we can talk of *when* there was no time, taking the 'when' in a non-temporal sense. It refers to eternity. Moreover, he gives a new account of eternity, 114,20-116,1, rejecting the conception of it as like a point. Since eternity is regarded as a measure of everlasting things, it had better have some kind of extension (*paratasis*), even though not a temporal one. We can imagine the extension there would be if the celestial clock were stopped.

These three issues are merely illustrations of Philoponus' original and densely packed replies.

*

A new introduction to the Commentators appears in R.R.K. Sorabji, *The Philosophy of the Commentators, 200-600 AD: A Sourcebook* (London & Ithaca, NY, 2004).

Notes

1. I have treated the infinity arguments in my *Time, Creation and the Continuum* (London & Ithaca, NY, 1983), ch. 14.

2. *Time, Creation and the Continuum* (London & Ithaca, NY, 1983), ch. 15, pp. 240-2.

Introduction

This translation is made from Rabe's 1899 Teubner edition,[1] the only modern critical edition of the Greek text. Departures from Rabe's text, many of which are based on Rabe's own suggestions in the critical apparatus, are mentioned in the notes as they occur and listed separately in front of the translation. Words in square brackets in the translation do not occur in the Greek but have been inserted to clarify the sense. Greek words are occasionally given in transliteration when it is thought their presence may help the reader.

The text is based on a single manuscript which is incomplete at either end and Proclus' first argument and part of Philoponus' response to it are missing. Fortunately, the missing argument (though not Philoponus' reply) survives in two Arabic versions and Peter Adamson has kindly prepared an English translation of one of them for this edition.[2]

Another consequence of the loss of the beginning of Philoponus' work is that we cannot be sure of its title or of that of the work of Proclus contained within it, and the various modern titles in currency are all based on the few, and not very helpful, ancient references to the two works and the conjectures of Renaissance scholars.[3] In English-language publications Philoponus' work is most frequently referred to by Rabe's Latin title *de Aeternitate Mundi contra Proclum* or by its English equivalent *Against Proclus On the Eternity of the World* and I have chosen to use the latter on the title-page of this volume.[4] In the remainder of this introduction and in the notes, on the other hand, I shall refer to it as *Aet.*, an abbreviated form of the Latin title. I have not needed to formulate a title for Proclus' work for the present volume, but Lang and Macro have recently edited it under the title *On the Eternity of the World.*[5]

In their introduction Lang and Macro argue (1) that Proclus' eighteen proofs are not aimed at the Christians but only at pagan philosophers like Plutarch and Atticus who took the creation myth in Plato's *Timaeus* literally, and (2) that, although Philoponus was a Christian, his quarrel with Proclus was on purely philosophical grounds and *Aet.* shows no evidence of his Christianity.[6] A reviewer[7] of Lang and Macro's work expressed unease with some aspects of their argument and the hope that these issues would be discussed at length in the introduction to this volume. A full discussion would go beyond the scope of this series,[8] but

I shall define my own position on the main points, since, as will become apparent, it has a bearing on some of my translation decisions.

I have no quarrel with the first part of Lang and Macro's argument. The issue that Proclus is addressing had a long history in Greek philosophy, going back at least to Plato and his immediate pupils, and is one that he also addresses in his commentary on Plato's *Timaeus*. The arguments he uses all have a long pedigree in pagan philosophy, many of them also appear in other works of his, and none of them has special application to the Christians. There is no overt reference to Christianity or the Christians in any of the proofs and, as mentioned earlier, I think it highly unlikely that there was any reference to the Christians in the title of the work.

None of this is surprising. Although he observed pagan religious practices, openly taught a philosophy that was in direct conflict with almost all aspects of Christian doctrine and probably came into direct conflict with the Christian authorities in Athens on at least one occasion, he does not openly attack, or even mention, Christians or Christianity anywhere in the considerable portion of his voluminous writings which survives.

On balance, this silence is more likely to have been the result of disdain or discretion, or a combination of the two, than of total indifference, and some scholars claim, I believe plausibly, to have found covert, discreetly hostile, references to Christianity in his works.[9] Are there any such in the eighteen proofs? Some[10] have seen one at 55,15-26, where Proclus writes:

> And so if someone, with the intention of paying reverence to him who is the cause of the universe, should say that he alone is everlasting while the world is not everlasting, he is, by denying that the latter is everlasting, declaring that the former is in movement and not unmoving. And if he says that he is in movement and not unmoving, he is saying that he is not always perfect but, because all movement is uncompleted activity, sometimes imperfect, and that he has need of something that is inferior [to himself] – I mean time of course – because of the very fact that he moves. And by saying that he is sometimes imperfect and not always perfect and that he has need of something inferior [to himself] he is being irreverent in the extreme. So if someone, with the intention of paying reverence to him who is the cause of the universe, should say that he alone is everlasting, he is being irreverent in the extreme.[11]

It seems likely that this is directed at a specific opponent or opponents and the Christians would certainly fit well enough. Lang and Macro suggest Plutarch or Atticus,[12] and although I can see nothing in the passages they cite[13] from Proclus and Plutarch to support their case, there is a passage in Proclus' eighteenth argument which suggests they are right. There, in an argument explicitly directed at Atticus and his associates,[14] he says:

> And most important of all, no one pays more respect to the cosmos than Plato, who also asserts that the cosmos is disordered, whenever god is not always uniform and unchanging, that is, whenever god is not an intelligible god; for the character of intelligible gods is to be uniform, unchanging and self-identical. So, either both are gods, both the cosmos and the demiurge, or neither; for disorder will make the one be not a god, while failing to be uniform and unchanging will make the other be not a god.[15]

In both passages the argument is at bottom the same: to downgrade the world is to downgrade God, and to pay due reverence to God one must pay due reverence to the world.[16] I find it difficult to believe that both arguments are not directed at the same opponents.

My guess is that after writing his commentary on the *Timaeus*, where he had addressed the same issues in what was in many ways a controversial work, he felt that there was room for a more systematic treatment of the same subject-matter, something more closely resembling the *Elements of Theology*. Of course, this does not mean that he was not aware that his arguments could be used against the Christians, or that a Christian reading them might be offended, or even that this might not have given him some pleasure, but the same could be said of much of what he wrote, and there is nothing in the proofs themselves to support the idea that they were aimed specifically at the Christians.[17]

It is the second part of Lang and Macro's argument that is problematical.

I shall begin by quoting a few passages that sum up their position.

> Philoponus too [sc. as well as Proclus] seems to have been writing within an entirely philosophical context, that is, Neoplatonism.[18]

> … there is virtually a complete absence of evidence for a Christian commitment in Philoponus' philosophical writings.[19]

> Christianity has left so little mark on his philosophy that his faith would not be known with anything like certainty – it might not even have been suggested – except for evidence from his theological writings.[20]

> … evidence for the presence and importance of Christian doctrine must be found in the *De Aeternitate Mundi contra Proclum* itself. And it is not there.[21]

> Philoponus indicates no Christian interests in his *De Aeternitate Mundi contra Proclum*; he does, however, indicate his commitment to Neoplatonism … Proclus' arguments and Philoponus' response to them play out in the arena of philosophy, not Christianity, and they do not present some supposed quarrel between Christianity and philosophy.[22]

I agree with some of this. Philoponus does write as a Neoplatonic insider and his arguments are strictly philosophical. Neither his stance in the creation debate nor most of his arguments were new to Greek philosophy and the bulk of the 646 pages of *Aet.* does indeed read like a contribution to an in-house philosophical debate.

However, a closer reading of *Aet.* shows that Philoponus frequently identifies himself as a Christian and his philosophical opponents as pagans, quotes from the Christian Bible, claims that Plato derived some of his positions from the Bible, and even seems to suggest that Philoponus believed that Proclus' arguments were directed specifically at the Christians.

Quotation of the key passages will be the best way to demonstrate this.

> I cannot help being amazed at how they brush aside arguments refuting [their own position] of which they are well aware, often even in the face of correctly-argued positions of their own, and taking up arms against the truth mislead those who are inexperienced in the subtleties of logical argument with every trick in the book.[23]

The 'truth' here is the Christian truth, as the phrase 'the truth of our Scriptures' in the passage quoted next, which refers back to this one, shows.

> Now that these facts have been demonstrated, how could one fail to be amazed, as I said at the outset,[24] at the trickiness of the great Proclus in argument? Although the facts speak so clearly for themselves, and although pretty well everyone accepts Aristotle's arguments … the philosopher takes no account of this – although no one would be so bold as to impute ignorance of such theories to him. [Instead] he has made it his one goal to arm himself by all available means against the truth of our Scriptures and, arguing against us as though we are novices in these matters, in the chapter before this has stitched together a fallacious argument … so that he can then draw the conclusion that it is either necessary that the divine should be in movement (that is to say, should alter) or, if that is impossible – for 'with God', as the Holy Scriptures somewhere[25] say, 'there is no change of position or shadow cast by

> turning' – that the world should have coexisted from everlasting with the being of God.[26]

These two passages show Philoponus identifying himself as a Christian and accusing his opponents, the pagan Neoplatonists[27] in general in the first passage ('they' can scarcely have any other reference) and Proclus in the second, of resorting to fallacious arguments in their eagerness to undermine Christian doctrine. The second passage could be seen as evidence that Philoponus believed that Proclus had aimed his proofs specifically at the Christians, but I am more inclined to think that Philoponus is actually attacking the teaching of anti-creationism in the Neoplatonic schools, of which Proclus' proofs were but one example, as an assault on Christian doctrine which could lead young, unsophisticated Christian students astray. It may even be that he had been such a student himself and that *Aet.* and the other creationist works which followed it were the result of his revulsion at what he saw happening after the scales had dropped from his eyes.[28]

Notice the quotation from the Christian Bible near the end of the second passage. Three more will occur in the passages I shall quote and there are seven in all in *Aet.*[29]

> I shall pass over the circle of Plutarch and Atticus, who are agreed by all to have explicitly affirmed that Plato believed the world to be generated with respect to time and to have taken issue with those of the contrary opinion. Past teachers of our gospel, among them Eusebius, who led the Church in Caesarea, have already cited lengthy extracts from them in their own works.[30]

Eusebius (d. circa 340) was a Christian polemicist and historian of the church who became the bishop of Caesarea in Palestine. The extracts from Plutarch and Atticus that are mentioned can be found in his *Praeparatio Evangelica* (*Preparation for the Gospel*), a polemical work in which he argues that Plato drew his inspiration from Moses and that his teaching helped pave the way for the Christian gospel. Philoponus shows evidence of having read the *Praeparatio* elsewhere in *Aet.*[31] His assumption that he can refer the reader to Christian writers for information about the views of Plutarch and Atticus shows that he is, if not exclusively, addressing a Christian audience.

> And this too [he has been discussing the myth in Plato's *Statesman*], as some of our [Christian writers] have correctly pointed out, he derived from the Holy Scriptures. For what else is 'assisting the revolution' of the heaven and then 'letting it go' 'when its circuits have fully completed the measure of time assigned to it'[32] than for 'the heavens to be rolled up and changed like a cloak?'[33] And listen to how Plato, once he had decided, again after hearing [the biblical

> words] 'for God did not make death', and 'he created all things that they might exist',[34] that the universe must remain immortal, states, again in this same *Statesman*, that immortality does not belong to the world by nature but comes to it newly acquired from the Creator: [Quotation follows].[35]

It is here that Philoponus claims that Plato was inspired by the Bible, and there may be an implication to the same effect at *Aet.* 142,8-11, where he writes:

> Only 'is', he says, should be said of God, stating this Mosaic [rule] quite explicitly. (For, appearing to the prophet, God declared 'I am He who is'[36]).

Philoponus may well of course have found the idea that Plato was inspired by the Bible in the *Praeparatio*.

Aet. was only the first of a series of either three or four books[37] in which Philoponus argued the case for a creation. None of the others survives in its own right, but we know quite a bit about them, largely because Simplicius attacks them at some length in his commentaries on Aristotle's *Physics* and *de Caelo*. The 'fragments' of the first of these, the *Against Aristotle On the Eternity of the World* (*de Aeternitate Mundi contra Aristotelem*) have been collected and translated by Christian Wildberg in this series[38] and fragments 132 and 134 contain clear references to the Christian expectation of a new heaven and a new earth.

Finally, Simplicius, who was Philoponus' contemporary, attacks him colourfully and viciously, often precisely as a Christian, and clearly had no doubt as to where his loyalties lay.[39]

I justified my excursion into these matters with the claim that my views have influenced some of my translation decisions. I particularly had in mind my treatment of the word *theos*. It is still common to translate it 'God' rather than 'god' when it occurs in a Christian author, whereas one is often torn between 'God', 'god' and 'the god' when it occurs in Plato or one of the Neoplatonists. In view of what has gone before, it will come as no surprise that I normally translate it 'God' when Philoponus uses it in one of his own arguments, and I have, for ease of application as much as anything, but also because I think he often reads *theos* as 'God'[40] in such cases, extended this to passages that he quotes from Plato, Aristotle, Proclus, and other authors. Although Proclus uses it a number of times in the eighteen proofs he does not, as it happens, use it in any of the proofs contained in this volume. I have not as a rule extended this capitalisation to pronouns or other words that refer to God.

This is probably the best place to say a few words on some of the other key vocabulary in *Aet.*, some of which appears as early as the title.[41]

In Plato's *Timaeus*, from which much of the terminology used in *Aet.*

and in the creation debate in general derives, the world, or universe, is variously referred to as *ho kosmos*, *ho ouranos* or *to pan*.

kosmos originally meant 'order' and, secondarily, 'adornment' and it never lost these connotations, but by Plato's day the meaning 'world-order' or simply 'world' was well-established. Common English equivalents are 'cosmos' and 'world' and I have opted for the latter.[42]

ouranos literally means 'heaven' but in the *Timaeus* Plato uses it interchangeably with *kosmos* (cf. *Tim.* 28B) and Aristotle at *Cael.* 278b ff. says that it may be used of (a) the outermost circumference of the universe (b) the heavens as a whole, including the stars, the sun, the moon and the planets (c) the universe as a whole. In *Aet.* it normally seems to be used in the second of Aristotle's three senses, but it is not always easy to see what is intended. I translate 'heaven'.

My rendering of *to pan*, which literally means 'the all', is 'the universe'.

I have thought it important to distinguish clearly between *aiônios* ('eternal'), *aïdios* ('everlasting') and *aei* ('always', 'for ever', etc.) in the translation. Proclus always reserves *aiônios* for entities which are outside of time, such as God or transcendental form, but uses *aïdios* or *aei* either of these same entities or of things which endure for ever in time, which, for him, include the world, matter, imminent form, generation and time itself. For Philoponus in *Aet.* things are a little more complicated. In reporting and refuting Proclus' arguments he observes the same distinctions; for example, he nowhere claims that Proclus is saying that the *kosmos* is *aiônios*. However, for him only eternal things are in fact *aïdios* and in one fairly lengthy passage (114,19-116,1) he can use *aiônios* and *aïdios* interchangeably to distinguish eternal things from those which exist in time. (In his earlier commentaries where he acted primarily as a reporter of Ammonius he was, of course, quite prepared to use *aïdios* of things which exist in time; at *in GC* 1,9-16, for example, he used it of the heavenly bodies and the four elements. Surprisingly, however, apart from two occasions in *Opif.*, he uses *aiônios* only in *Aet.*).[43]

Using 'everlasting' for *aïdios* entails using 'everlastingness' for *aïdiotês* (although, as mentioned earlier in this introduction, I retain the by now traditional 'eternity' in the volume title) and I have even thought it best to use the unlovely coinage 'co-everlasting' rather than 'co-eternal' to translate *sunaïdios*.

Like Plato in the *Timaeus*, Proclus and Philoponus most commonly refer to the maker of the *kosmos* as *ho theos* or *ho dêmiourgos* (although Proclus does not use the former before Argument 8). Common translations of *dêmiourgos* are 'demiurge', 'craftsman' (a more or less literal rendering of one of the senses of the Greek word) and 'creator'. In translating the *Timaeus* I would use either 'demiurge' or 'craftsman', but in Proclus and Philoponus the term has, I think, lost much of its original force and I have opted for 'creator', partly because doing so makes it easier to find English equivalents for the related words *dêmiourgein* ('to create'), *dêmiourgêma* ('a creation'), *dêmiourgia*, ('creation') and *dêmiourgikos* ('creative'). I have already discussed the translation of *ho theos*.

Both creationists and anti-creationists were eager to enrol Plato on their side and Philoponus' debate with Proclus and other creationists is in part over the correct interpretation of the *Timaeus*. In this debate one of the key issues was the correct interpretation of the verb *ginesthai* and related words when applied to the *kosmos*.

The LSJ article on *ginesthai* (which is there listed under its earlier spelling *gignesthai*) is organised into two main sections. The first is headed 'abs. [sc. without a predicate], *come into being*', and includes, amongst others, subsections headed '*to be born*', and '*to be produced*'. The second is headed 'folld. by a Predicate, *come into* a certain state, *become*, and (in past tenses) *to be*'. There is no doubt that Plato often uses *ginesthai* of the physical world to express the idea that it is in perpetual flux, always changing and 'becoming' different (a usage which, although he commonly uses the verb *without* a predicate, would fall under LSJ II), and this is not a matter of dispute between Proclus and Philoponus (see especially VI,15-16). The question at issue between them is whether he also applies the verb to the *kosmos* as a whole in a sense that would fall under LSJ I. Philoponus claims that he does, at *Tim.* 28B for example, where he understands *gegonen* in the sense 'it has come into being' (or perhaps even 'it has been generated', or 'it has been created'), while Proclus argues that he does not, unless perhaps in a very attenuated sense. This, of course, means that the same words will often mean something different to Proclus and Philoponus, which makes life difficult for the translator. One popular solution, which I shall adopt, is to translate *ginesthai* 'to come to be', which can, with charity, be understood as embracing both 'to come into being' and 'to become', and as therefore adequately covering most relevant senses of *ginesthai*. (This only applies to contexts where the creation of the *kosmos* is at issue. *ginesthai* is something of a portmanteau word and I translate it in many different ways in other contexts).

The choice of 'come to be' for *ginesthai* raises the possibility of something like 'coming-to-be' and 'admitting of coming-to-be' (or, on a different view of the word, 'having-come to-be') for the related words *genesis* and *genêtos*, as used by Hussey in his translation of the third and fourth books of Aristotle's *Physics*.[44] However, the two words are both very common in *Aet.* and in some passages, especially in the case of *genêtos*, this would become intolerably cumbersome, so I have, rather illogically, opted for 'generation' for *genesis* and 'generated' for *genêtos*.

The choice of 'generated' for *genêtos* raises another issue. Verbal adjectives in *-tos* (of which *genêtos* is one) may express possibility or have the force of a perfect passive participle. Some display only one of these possibilities, others, including *genêtos*, as the entries in LSJ and

Lampe when taken together show, both. Of course, when an adjective is capable of either signification, it is not always clear which is intended. In fact, one suspects that the writer would often not have found it easy to say. This being so, it is not surprising that *genêtos* in *Aet.* has been read either way. To take only two examples, Dillon,[45] when translating the excerpts from Taurus in *Aet.* VI, renders it 'created', while Judson, in his article on generability and perishability in Philoponus,[46] prefers 'generable'. My own view is that the 'perfect passive' sense of the word is usually uppermost in the minds of both Proclus and Philoponus and it is for that reason that I have preferred 'generated' to 'generable'. (In fact, it seems to me that in VI,9, in the course of dismissing the first of the various meanings that Taurus had proposed for *genêtos*, Philoponus comes close to rejecting the meaning 'generable' altogether).

In philosophical texts, including *Aet.*, *ginesthai*, *genesis* and *genêtos* are often opposed to *phtheiresthai*, *phthora* and *phthartos*. *phtheiresthai* is the passive of *phtheirein* ('to destroy') and so can be rendered 'to be destroyed', but 'to perish', 'to pass away' and 'to cease to be' are all commoner. I usually (but not, as the Greek-English Index shows, always) use 'perish' for *phtheiresthai* and 'perishable' for *phthartos*. For *phthora*, because 'perishing' does not always work well, I usually use 'passing out of existence'.[47]

I discuss the translation of a number of other words in the notes, usually at their first occurrence.

The early stages of the preparation of this translation were supported by a grant from the Australian Research Council. I would like to thank Frans de Haas, James Wilberding, and Andrea Falcon who each read part of a draft of the translation and made many valuable suggestions, the Ancient Commentators on Aristotle Project editorial team for assistance, patience and support, and Richard Sorabji for his advice and encouragement.

Notes

1. H. Rabe (ed.), *Ioannes Philoponus de Aeternitate Mundi contra Proclum*, Leipzig, 1899 (reprinted Hildesheim, etc., 1984).

2. For details of the version translated by Adamson, see his notes; for the other version, which is known from two manuscripts and is apparently an earlier and less sophisticated translation of the same Greek text, see F. Rosenthal, 'From Arabic books and manuscripts VII: some Graeco-Arabica in Istanbul', *Journal of the American Oriental Society* 81 (1961), 9-10. Although mediaeval Arabic bio-bibliographers refer to an Arabic translation of Philoponus (see G. Anawati, 'Un fragment perdu du *De aeternitate mundi* de Proclus', in *Mélanges de philosophie grecque offerts a Mgr. Diès* (Paris, 1956), 21-2; R. Wisnovsky, 'Yaḫyâ al-Naḫwî', *Encyclopaedia of Islam*, vol. 11 (Leiden, 2002), 252), it does not seem to have survived.

3. Philoponus himself refers to the present work at least half a dozen times in later works. At *Opif.* 88,21 he calls it *ta eis ta Proklou graphenta hêmin* ('what has been written by us against the [writings] of Proclus'); in his *Against*

Aristotle, as cited at Simplicius *in Cael.* 135,27, *hoi elenkhoi pros Proklon* ('the refutations against Proclus'); and in the *Against Aristotle*, as cited at Simplicius *in Cael.* 136,17 and Simplicius *in Phys.* 1141,9; 1142,1 and 1159,2, *ta pros Proklou* ('the [writings] against Proclus'). The article on Philoponus in the *Suda*, a tenth-century encyclopaedia, says that he wrote *kata tôn dekaoktô Prokleiôn epikheirêmatôn* ('against the eighteen Proclean proofs'), and the article on Proclus includes *Epikheirêmata kata Khristianôn iê* ('*Eighteen Proofs Against the Christians*') in a list of his works and later in the article states: 'This is the Proclus who, second only to Porphyry, used his foul and insolent tongue against the Christians. John Philoponus wrote against him, replying most brilliantly to his eighteen proofs (*tôn i kai ê epikheirêmatôn*) and showing that even for a Hellene, amongst whom he had a high reputation, he was uncultured and stupid'.

In all of this only *Eighteen Proofs Against the Christians* looks like a formal citation and I have to agree with Saffrey (H.-D. Saffrey, 'Allusions antichrétiennes chez Proclus: Le Diadoque platonicien', *Revue des sciences philosophiques et théologiques* 59 (1975), 553-63) that this looks like an invention based on the fact that Proclus' proofs, which are not openly anti-Christian and are therefore unlikely to have included the words 'Against the Christians' in their title, were refuted by the Christian Philoponus. It is not until the fifteenth century, when someone wrote the words *Iôannou Alexandreôs tou Philoponou kata tôn Lukiou Philosophou Proklou peri Aïdiotêtos Kephalaiôn* ('John Philoponus of Alexandria's Against the Lycian Philosopher Proclus' Chapters on Eternity') and someone else *estin Iôannou Alexandreôs tou Philoponou hê parousa kata aïdiotêtos pragmateia aristê tôn peri toutou* … ('The present work on eternity, the best on the subject, belongs to John Philoponus of Alexandria …) in the main manuscript of Philoponus, which itself dates from the ninth or tenth century, that we meet a reference to Philoponus' work containing the word 'eternity', and not until the sixteenth, when the first printed edition and a late copy of the principal manuscript both call it *kata Proklou peri Aïdiotêtos Kosmou* ('*Against Proclus on the Eternity of the World*') that we find a title containing the phrase 'the eternity of the world'.

4. Because, for reasons I shall explain later, I translate the Greek word *aïdiotês* 'everlastingness' rather than 'eternity', I translate the (rather different) Greek title that Rabe supplies at the beginning of the Greek text: '*Against the Arguments of Proclus Concerning the Everlastingness of the World*'.

5. H. Lang and L. Macro (eds.), *On the Eternity of the World* (*De Aeternitate Mundi*), *Proclus* (Berkeley, etc., 2001). Their edition includes a Greek text based on Rabe's, with their translation of it; a text and translation by Jon McGinnis of the Arabic version of Proclus' first proof that is translated in this volume; the Latin version of Proclus' proofs from the earlier of the two surviving sixteenth-century Latin translations of *Aet.*; a substantial introduction and useful notes. The seventeen proofs of Proclus which survive in *Aet.* have also been translated into English by the English Neoplatonist Thomas Taylor in *The Fragments That Remain of the Lost Writings of Proclus, Surnamed the Platonic Successor* (San Diego, 1988) [originally published 1825], 35-92, and into German in M. Baltes, *Die Weltentstehung des platonischen Timaios nach den antiken Interpreten* (Leiden, 1976), vol. 2, 134-64. Only small portions of Philoponus' refutations of them have been translated into any modern language.

6. op. cit., 4-14.

7. Dirk Baltzly, *Bryn Mawr Classical Reviews*, BMCR 2002.10.19.

8. Which aims to provide accurate, readable translations with a minimum of annotation.

9. See especially Saffrey, 'Allusions antichrétiennes chez Proclus', which Lang and Macro cite in nn. 15 and 20 to the Introduction. Although Lang and Macro are initially rather ambivalent about the existence of such references, they seem to come down in favour of it in the last paragraph of p. 7.

10. For example Baltes in *Die Weltentstehung des platonischen Timaios nach den antiken Interpreten*, 2,139, who cites P. Bastid, *Proclus et le crépuscule de la pensée grecque* (Paris, 1969).

11. My translation.

12. op. cit., 7 and 52, n. 10.

13. ibid., 52, n. 10.

14. At *Aet.* 606,17.

15. *Aet.* 608,11-18 in Lang and Macro's translation.

16. Lang and Macro's summary of the second passage (op. cit., 139) shows that they read the argument as I do.

17. Nor, incidentally, do I think, as Lang and Macro do (ibid., last paragraph on p. 7), that they were primarily intended as a refutation of the long-dead Plutarch and Atticus. Atticus is mentioned only once and Plutarch not at all, and although one or both of them may be in the offing from time to time on other occasions, most of Proclus' arguments have no direct bearing on anything they wrote. They were adequately dealt with in the commentary on the *Timaeus*.

18. ibid., 7.

19. ibid., 10.

20. ibid., 11.

21. ibid.

22. ibid., 12 (with omission).

23. *Aet.* 61,5-9.

24. The reference is to the passage just quoted.

25. *James* 1,17.

26. *Aet.* 74,24-75,21 (with omissions).

27. Philoponus frequently refers to the pagan philosophers, or the pagans in general, as 'Hellenes' or 'children of the Hellenes' (a phrase which occurs in Eusebius), a practice that was common among both Jews and Christians. (For instances see *Aet.* 37,13; 241,14; 245,23, etc.).

28. Of course, it is easy to imagine other possible reasons, some of them less creditable, for Philoponus' attack on Proclus. There is a good survey of those that have been advanced in K. Verrycken, 'The development of Philoponus' thought and its chronology', in R. Sorabji (ed.), *Aristotle Transformed: the Ancient Commentators and Their Influence* (Ithaca, NY, 1990), 234-6 (modern views) and 258-60 (Arabic writers and Simplicius). There was a tradition among Arabic philosophers that his anti-eternalist works did not represent his true convictions and that his fellow Christians either bullied him into writing them or paid him to produce them. Modern views range from a conversion from paganism to Christianity to a reaction to Justinian's closure of the Neoplatonic school in Athens, motivated either by personal ambition or a desire to deflect the Emperor's attention from Alexandria.

29. The three I do not quote can be found at *Aet.* 6,6; 6,7 and 128,15.

30. *Aet.* 211,10-18.

31. Perhaps the clearest instance is at 37,13-15, where the words: 'Among the Hellenes (*par' Hellêsi*) indeed shrines and oracles [where] an evil demon feigned the true foreknowledge of God were established in every corner of the land' seem to be a clear echo of the last part of the last sentence of subsection 5.3.10 of the *Praeparatio*, which reads: 'On the shrines and oracles among the gentiles (*para tois ethnesi*) belonging to evil demons'. (*para tois ethnesi* and *par' Hellêsi* have much the same reference in a Christian context).

32. Plato, *Pol.* 269C.

33. A paraphrase of *Hebrews* 1,12, which is itself an adaptation of *Psalms* 102, 27.

34. *Wisdom of Solomon* 1, 13 and 14.

35. *Aet.* 229,9-21.

36. *Exodus* 3,14.

37. On these see C. Wildberg (tr.), 'Simplicius: Against Philoponus on the Eternity of the World', in *Place, Void, and Eternity* (Ithaca, NY, 1991), 100.

38. C. Wildberg (tr.), *Philoponus: Against Aristotle on the Eternity of the World* (Ithaca, NY, 1987). In his 'Prolegomena to the Study of Philoponus' *contra Aristotelem*', in R. Sorabji (ed.), *Philoponus and the Rejection of Aristotelian Science* (London & Ithaca, NY, 1987) Wildberg argues that the *Against Aristotle* ended with two or more books dealing, at least in part, with Christian eschatology.

39. See Philippe Hoffmann, 'Simplicius' Polemics: Some Aspects of Simplicius' Polemical Writings Against John Philoponus: From Invective to a Reaffirmation of the Transcendency of the Heavens', ch. 3 of R. Sorabji (ed.), *Philoponus and the Rejection of Aristotelian Science* (London & Ithaca, NY, 1987). In Simplicius, as in earlier Neoplatonists, many of the references to Christianity are 'coded', but some are quite open (the reference to the Psalms, for example, and the use to which Simplicius puts it, which are mentioned by Hoffmann on pp. 70-1). Simplicius could be bolder than his predecessors because his commentaries were not produced in a teaching context.

40. In fact, although he is well aware of the complexities of the Neoplatonic system, (for evidence of this in *Aet.*, see 90, 24ff.), he is normally able to sidestep them and write as though he were debating creation with a fellow monotheist. This is possible (and even reasonable) because Proclus himself writes in the abstract, metaphysical manner of his *Elements of Theology* rather than from the theological perspective of his commentary on the *Timaeus* or the *Platonic Theology*, so that it scarcely matters, to take an example, that the *dêmiourgos* is one of a triad of gods in the second hypostasis for Proclus whereas for Philoponus he is the Christian God.

41. Lang and Macro have a useful section describing some of the difficulties of translating Proclus and documenting some of their translation decisions. (op. cit., 28-33).

42. For more on the history and meaning of *kosmos*, see W.K.C. Guthrie, *A History of Greek Philosophy*, vol. 1 (Cambridge, 1967), 110-11 and 208, n. 1.

43. For an overview of the use of *aïdios*, *aei* and other time words in antiquity see R. Sorabji, *Time, Creation and the Continuum*, 112-17. Proclus elsewhere explicitly distinguishes temporal and non-temporal uses of *aïdios* and *aei* and Philoponus too finds room for a non-temporal use of *aei* both in *Aet.* (104-7) and in other works; references ibid., 115, nn. 66, 67 and 74.

44. E. Hussey (tr.), *Aristotle's* Physics *Books III and IV* (Oxford, 1983).

45. J. Dillon, *The Middle Platonists: a Study of Platonism, 80 BC to AD 220* (London, 1977), 242-3.

46. L. Judson, 'God or Nature? Philoponus on Generability and Perishability', in R. Sorabji (ed.), *Philoponus and the Rejection of Aristotelian Science* (London & Ithaca, NY, 1987).

47. There is a good discussion of possible translations of *ginesthai* and *phtheiresthai* in the 'Introduction' to C. Williams, *Aristotle's* de Generatione et Corruptione (Oxford, 1982).

Departures from Rabe's Text

Emendations other than my own are credited. Those attributed to Rabe are based on suggestions printed in Rabe's apparatus. All departures from Rabe's text are also recorded in the footnotes, in the case of my own emendations often with a brief justification. With the exception of his restorations at the beginning of the text, where I reproduce his angle brackets, I do not indicate Rabe's own departures from the manuscript tradition either in the translation or in the notes.

2,8	Adding *hupo khronon dê kai ho* after *khronon* (Rabe).
6,14	Changing *paragesthai* to *paragetai* (Rabe).
10,19	Adding *legô dê* after *adunaton* (Rabe).
11,3	Adding *anthrôpôn* before *tôn* (Rabe).
11,6	Adding *ei* after *kai.*
14,20	Changing *proüparkhon* to *proüparkhoi* (Rabe).
16,18	Changing *ekeinôn* to *ekeinou* (Rabe).
16,25	Deleting *epeidê to phôs ditton estin.*
16,26	Deleting *sunupostan.*
18,26	Changing *hupo* to *epi* (Rabe).
24,10	Adding *kai to paradeigma ouk ên* after *hote ouk ên.*
24,11	Deleting *kai to paradeigma.*
27,20	Deleting *aei.*
28,18	Changing *genomena* to *ginomena* (Rabe).
28,23	Deleting *einai* (Rabe).
29,10	Changing *parakousantas* to *epakousantas* (Rabe).
31,8	Adding *kai* before *diapherontôs* and *en* after it (Rabe).
33,2	Changing *huphestêken* to *sunestêken* (Rabe).
33,16	Deleting *allou deomenon eis huparxin* (Rabe).
33,17	Adding *deomenon eis huparxin* after *allou* (Rabe).
35,28	Repositioning *ousias* to follow *legô* (inside the bracket).
36,14	Adding *allôn tôn* after *tôn* (Rabe).
39,17	Deleting *hen.*
40,12	Changing *touto* to *hoper.*
40,22	Deleting *hen.*
43,14	Adding *ginesthai* after *energeiâi* and changing *aitiou* to *aitiôi.*

43,22	Deleting *heteron* (Rabe).
44,12	Adding *to* after *hoti* (Rabe).
44,21	Deleting *energeiâi*.
44,24	Changing *ton* to *to* (Rabe).
44,25	Changing *autos* to *auto* (Rabe).
45,12	Changing *onta* to *gnonta* (Rabe).
45,16	Adding *tou* before *energeiâi* (Rabe).
45,17	Adding *tou* before *dunamei* (Rabe).
45,18	Adding *tou* before *energeiâi* (Rabe).
46,19	Adding *auta dekhesthai* after *pephukenai*.
46,26-47,1	Changing *elegeto, ê ho* to *elegeto einai, ho* (Rabe).
47,14	Changing *epitêdeiotêta* to *dunamin*.
48,1	Changing *ouketi* to *ouk esti* (Rabe).
49,18	Changing *tôn* to *tou* (Rabe).
50,10	Adding *to* before *axiôma* (Rabe).
51,12	Changing *tôn* to *tou* (Rabe).
51,24	Changing *autos* to *auto*.
52,9	Changing *houtos* to *houtôs*.
54,19	Adding *to* before *dunamei* and changing *auto* to *autou*.
54,25	Adding *autos heautôi aitios ên tês ek tou proterou dunamei eis to energeiâi metabolês* after *autês*.
56,10	Changing *aïdiou* to *aïdion*.
57,15	Deleting *ou* (Rabe).
57,21	Changing *oude* to *ou dia*.
59,4	Changing *atelês ôn* to *atelê*.
66,18	Punctuating with a comma rather than a full stop after *akhronos*.
66,25	Punctuating with a full stop rather than a comma after *kinêsis*.
70,11	Changing *tois* to *hois* and repositioning *pôs* to follow *adioristôs*.
71,15	Adding *ho* after *hoti*.
72,11-12	Changing *kai thermou de homoiôs kai psukhrou* to *kai ho thermainomenos de homoiôs ek psukhrou*.
74,26	Changing *orthotêtos* to *deinotêtos*.
79,26	Changing *eis to autou tou Sôkratous sôma* to *eis to auto tôi tou Sôkratous sômati*.
81,20-1	Changing *ho autos nous hôste tas* to *ho autos anthrôpos nosôn te tas* (Rabe).
81,21	Adding *ê* after *hugiainôn*.
82,6	Changing *dêmiourgêsei* to *dêmiourgêsai* (Rabe).
82,15	Removing the existing diacritics from *tauta* and accenting it with a circumflex on the first syllable.
89,14	Changing *ekeino* to *ekeina*.
90,5	Changing *aitiôtaton* to *aitiôteron* (Rabe).
90,10	Changing *eti* to *estin*.

90,11-12	Changing *tôi aitiôi tês kinêseôs estin* to *tôi aitiôteran tês kinêseôs einai.*
92,5	Adding *aei* before *estin* (Rabe).
92,18	Deleting *dunamei kai* after *deuteros.*
92,20	Changing *ê* to *ei.*
92,22	Changing *ê* to *eiê.*
95,7	Adding *dedêmiourgêtai kai aei* after *aei* (Rabe).
96,5	Changing *einai* to *esti* and closing the brackets after it rather than after *thraxeien* at 96,4.
96,6	Changing the full stop after *hêmin* to a comma.
96,7-9	Removing the brackets around *hoson ... meros* and punctuating instead with a semicolon after *estai* and a full stop after *meros.*
96,19	Changing *dia ti* to *dioti.*
96,21	Punctuating with a comma rather than a question mark.
98,11	Changing *pote* to *ton te.*
99,19	Changing *genesthai* to *ginesthai* (Rabe).
100,18	Changing *anankê* to *anankazei.*
101,2-3	Changing *tou ... parêgmenou* to *tôn ... parêgmenôn.*
101,6	Changing *hôsper* to *eiper.*
103,25	Adding *aei* after *estin* (Taylor and Rabe).
104,16	Changing *tina huparxin* to *tinos huparxeôs.*
104,27	Changing the second *ên* to *êi* (Rabe).
107,21	Changing *pausetai* to *pauetai* (Rabe).
111,27	Adding *gar* after *ean* (Rabe).
112,3	Deleting *ouk.*
112,5	Deleting *ouk.*
112,17	Adding *ti* after *atopon* (Rabe).
113,18	Deleting *kai* after *tmêmata* (Rabe).
115,2	Changing *temnomenos* to *temnomenon* (Rabe).
115,17	Adding *kai kata to paradeigma tês aiôniou phuseôs* after *ginêtai.*
115,22	Adding *ton aiôna* after *zôês.*
116,14	Changing *khronikon* to *khronikôs* (Rabe).
117,19	Adding *legonta* after *Platôna* (Rabe).
117,27	Changing *ex aiônos* to *exô henos* (Aristotle).
118,17	Changing *dei* to *dein* (Rabe).
118,19	Adding *logon* after *toiouton* (Rabe).

PHILOPONUS

Against Proclus
On the Eternity of
the World 1-5

Translation

<John Philoponus the Alexandrian's Against the Arguments of Proclus Concerning the Everlastingness of the World>[1]

<Preface>
<The First Argument[2] of Proclus the Successor>[3]

The first argument of Proclus to demonstrate that the world is 34,3 everlasting.[4] He says that the first of the arguments we use to show that the world is eternal is drawn from the generosity of the Creator. No more convincing demonstration can be given than the one based on the fact that the universe[5] is like the One Who gave it reality, and from Whom is its existence. For, because the coming-to-be of the universe is from His generosity[6] alone, [His generosity] produces it, since one may not say that He creates it through anything other than generosity without admitting that He is generous at one time but not at another. He is always the cause of the existence of the world, since the being[7] of the world is equal[8] to the being (*kawn*) of the Creator. We do not find anything at all to which it belongs that it makes the world out of generosity, yet does not make it everlastingly; [for] He is everlastingly generous. Since He is everlastingly generous, He everlastingly wants[9] all things to be like Himself. If He wants all things to be like Himself, then He is able to make all things like Himself, since He is the ruler and lord of all things. But if He wants to make all things like Himself, and He is able to make all things like Himself, then He makes them everlastingly. For, everything that does not make refrains from making either because it does not wish to make or because it is unable to make – if it is susceptible to one of these two. So if the Creator – may He be exalted[10] – makes the world from His generosity, He makes it everlastingly.

Thus it is necessary that the world did not come to be at some time in the past, and that it will not perish at some time in the future.[11] For the claim that He is unable to do what He wishes deserves ridicule, because if He were able sometimes and not able at others, 35,1 then it would follow that He admits of change and affection. For His losing the ability would be the reason for His receiving an affection. Whatever alters from not being able to being able has changed, because both power[12] and lack of power belong to [the category of] quality, and change is alteration of quality. If He is everlastingly able to create, and everlastingly wishes to create, then it is absolutely necessary that He everlastingly creates and that the universe is everlastingly created, and that the world is everlastingly existent (*mawjûdan*), just as the Creator[13] is everlastingly creating.

However, the Creator is eternally existent, but the world eternally comes to be. The meaning of 'everlastingly' is not one and the same for both, but rather for the Creator its meaning is timelessness and eternity. But for the world its meaning is time that is infinite, because what is paired with[14] existence is timelessness and eternity, but what is paired with coming-to-be is time.

<The Sections of the Refutation of the First Argument>

...

<The Refutation of the First Argument>

1. ...

2.[15] ... to equate what exists eternally and what [exists] indefinitely in time; for he says that eternity and time are not the same.[16]

So let us take it as already proved by this that in his view too eternity is infinite power.[17] So if, according to both Proclus and Aristotle, eternity is infinite power, and if, in a word, that which is
2,1 eternal is in every case also infinitely powerful, then whatever does not partake of infinite power does not partake of eternity, and whatever does not even partake of eternity is not eternal. So if the divine does not partake of infinite power under either of the two above-mentioned hypotheses, it will not be eternal under either of them, and if it is not eternal, it is of necessity temporal; for of all things some are eternal, others temporal.[18] So even the creator of time is temporal![19] And if, since it is not infinitely powerful, the divine is not eternal, nor [for the same reason] will it be everlasting;[20] for if [its] power is finite, it cannot last for ever. Therefore either it will have a beginning and an end because of the finite and non-eternal [nature] of its power, or there will be some other cause of being and everlastingness for it. So let us once more raise the same puzzles about this [cause] as well. For one must enquire concerning it too whether or not it could bring into existence things that are superior to those it [actually] brings into existence, since, if it brought the other [god] into existence and gave
3,1 it the ability to create, it too has creative power. And one must either continue this absurdity *ad infinitum* or suppose that the creative power of God is infinite, or rather that the divine is in *all* respects infinitely powerful.

Moreover, if it is agreed that God has knowledge of all things – of those there have been, those there are, and those there will be – but the world is, as they hold, without beginning and without end, and for that reason the things that have been and that will be are infinite, it

is not possible for God, if his cognitive power is finite and not infinite, to fully grasp knowledge either of what has been or of what will be. For it is impossible to master what is infinite with finite power. And 4,1
so God will have knowledge neither of all that has been nor of all that will be. For if he is going to have [such knowledge], either the world will not be ungenerated and imperishable, so that what has been and what will be would not be infinite, or it is not by means of a finite power that God knows things. And if this is so, nor does he create by means of a finite [power].

If, then, when the two above-mentioned hypotheses are advanced ([1] the one that states that the power of God is one and simple[21] and that God creates by means of a finite power, and [2] the one that hypothesises that God has many powers, but none of them infinite), the most absurd consequences follow ([namely,] [a] that God is not eternal but temporal, and [b] that he needs another [god] to furnish him with everlasting (*eis aei*) permanence, although *he* does not escape the same absurdity either, and, further, if it is agreed that the world is, as they hold, ungenerated and imperishable, [c] that God has knowledge neither of what has been nor of what will be), then neither of these hypotheses will be true.

Three hypotheses still remain: either [3] the power of God is one and he effects and produces all things by means of infinite power (whether it is a question of his creating, of his exercising providence, of his knowing, or of anything else); or [4] his powers are many and each of them infinite; or [5] they are many and some of them infinite, some finite.

Now, should some be infinite and others finite [5], [any] account 5,1
will be arbitrary and full of caprice. For on what authority is it possible to say that this power of God is infinite and that finite? For the divine will be in part eternal and in part temporal and the very power of God that brings time into existence will be temporal. And how will he maintain things for ever if it is not infinite? For one presumably in every case maintains what one creates by the [same] power with which one creates it. So the third hypothesis too is shown to be absurd and impossible.

It remains, then, than one of the remaining two is true, I mean that either [4] the powers of God are many and each of them infinite, or that [3] [his power] is one and simple and the whole of God's activity takes place with infinite power.

The latter [alternative] [sc. 3] is, because of the perfect simplicity of God, both more true and consistent with [our] common conceptions of God.

3. It is clear, then, from what has been said that it is impossible that God creates things with [merely] finite power. And if it is impossible for God's creative power to be finite, it will of necessity be infinite. (I mean infinite not only in the sense of always being able to

bring [things] into existence – for it is impossible for a power that is not infinite in this sense to last in perpetuity – but also, if I may put it so, that at each instant the production of each and every thing, for
6,1 example of the sun, of soul, of the heaven, by absolutely infinite power is taking place).

A proof of this in addition to those that have already been stated would be the fact that God brings all things into existence timelessly and together with the thought.[22] [Our] common conceptions teach us that all believe this of God. Thus of our own prophets one[23] says: 'He spoke and they came to be', and another:[24] 'You had the thought and they came to be; you minded it and they were present'. For he who brings even time itself into existence assuredly would not have need of temporal extension to produce things. And is it not indisputable proof of [his] infinite power that God's thought that he wished them to come to be sufficed on its own for the bringing into being of all things?[25]

Moreover, this being so, it is by infinite power that both the greatest and the least of things are brought into existence.[26] For if he causes all things to exist by one power, as even Proclus believes – for in his work *On Ten Puzzles About Providence*[27] he says that 'the unitary knowledge of providence is, in the same undivided [entity], the knowledge of all divided things, both of all that are most individual and of all that are most universal; and just as it has caused each thing to exist in unity (*kath' hen*)', he says, 'so does it know each in unity' – so if [as I was saying] he causes all things to exist in unity, and [his] power is not parcelled out among or completed by things that come to be, and if things that have come to be are various, some being superior, some inferior, and if, as has been shown, he causes all things to exist by infinite power, then he has produced both the
7,1 greatest of things and the least by one and the same infinite power.

Since, then, it has been shown that it is impossible for God's creative power to be finite and that it is necessarily infinite, it is not true that it is either on account of the productive cause alone or on account of both [factors together] – that which produces and that which comes to be – that neither more nor better things than those which exist have come to be; for on [either of] these hypotheses it would follow that God's creative power is finite, which has been shown to be both absurd and impossible.

It therefore remains that the only hypothesis left is true – the one that states that the nature of things that are generated is the sole reason for the impossibility of creating more or better things. And even if one admits that both [factors], I mean that which produces and that which comes to be, are causes of the impossibility of creating more or better things, it is in every case necessary that it be for the self same reason that a thing (*to ginomenon*) cannot come to be and that that which produces cannot produce [it].

So let the reason that one would give for the impossibility of more or better things coming to be than those that have come to be even though, as has been demonstrated, God's creative power is in all respects infinite, be also accepted by us as the reason for the impossibility of the world having existed from everlasting even though God has always been good and always a creator. And the most scientific[28] explanation of these things that we are able to offer is as follows.

Aristotle has proved by many arguments in the *Physics*[29] and in the *On the Heaven*[30] that the infinite can in no way actually exist, and this is agreed to be so by all without exception.[31] (By infinite I here mean [infinite] in magnitude as regards extension or in number; there is every necessity that the infinite in power should exist in things that are everlasting by nature). And since the infinite in magnitude or in number is not of a nature to exist in actuality, it is absolutely necessary that the number of things be finite. And this being so, it is ridiculous to ask why more things do not exist; for, given that the infinite cannot emerge in actuality, the same puzzle will still remain, no matter how many more things one hypothesises to exist in addition to the things that [currently] exist. So in this way it is necessary that the number of things be finite.

And, again, things superior to those that exist have not come to be because it is absolutely necessary that things brought into existence by the Creator be inferior to him. So the more superior are the things whose existence one hypothesises to those that [currently] exist, so much the more, since they [nevertheless] fall short of the inconceivable power of God, would one still leave the same problem [unsolved]. And so it is necessary that the quality of created things too be circumscribed, just as their number is.

So if the fact that more or better things are not created has its cause in the nature of created things and this in no way detracts from the power or goodness of God, then we, by hypothesising consistently with this that the world is not ungenerated, neither accuse the creator of weakness nor [claim] that he does not always wish to bring good things into existence, but find the cause in the nature of created things. For just as it follows from the fact that the infinite cannot actually exist that more, or an infinite number of things are not created, so too do I say, again so as not to allow that the infinite actually exists and make the infinite traversable, that the world cannot coexist with God from everlasting. If the world *had* existed from everlasting, it would be absolutely necessary for the number of things that have come into existence in the world from the beginning up until now – I mean men and plants and the other individuals in each species – to have become actually infinite as well. For should one hypothesise that the number of men or plants or of individuals of any other kind that have come to be is finite, since each of them has had its existence in a finite time, it would also be necessary for the whole of time to be finite; for that which consists of finite [parts] is finite.[32] So since, if the world is ungenerated, the time that has elapsed is also actually infinite, the individual things that have come to be in this infinite time must, I imagine, be actually infinite in number too. And so it will follow that, if the world is ungenerated, an infinite number actually exists and has occurred. But that the infinite cannot in any way exist in actuality, whether by existing all at once or by coming into existence bit by bit, we shall, if God permits, in due course demonstrate at greater length in another [work],[33] when, having looked into all of the puzzles surrounding the issue of an everlasting world,[34] we shall finally establish on our own account that it cannot be everlasting. And I shall cite Aristotle himself explicitly arguing this very point – I mean that the infinite can in no manner exist in actuality. But I believe that this is in fact self-evident. For the very same reason (*aitia*) that the infinite cannot exist all at once and at the same time it cannot emerge into actuality by existing a bit at a time. For if it were at all possible for the infinite to have emerged into actuality by existing a bit at a time, what further reason (*logos*) could there be to prevent it from also existing in actuality all at once? For it would seem much more impossible to claim that the infinite is brought to actual birth bit by bit and, as it were, counted out one unit after another than that it exists all at once and at the same time. For, if it exists all at once, perhaps there will be no need to go through it unit by unit and, as it were, count it off;[35] but if it comes to be a bit at a time and one unit always exists after another, so that eventually an actually infinite number of units has come to exist, even if it does not exist all at once at the same time because parts of it have ceased to exist while parts [still] exist, it has nevertheless become traversable. [But] this – I mean[36] the traversing of the infinite by, as it were, counting it off unit by unit – is impossible, even if the counter were everlasting. For the infinite is by its nature untraversable; otherwise it would not be infinite. So if the infinite is untraversable, but successive generations of the [human] race, advancing individual by individual, have descended through an infinite number of individuals to those that exist now, then the infinite has become traversable, which is impossible. Therefore the number of earlier individuals is not infinite. For [if it were] the generations of the race would not have reached down to each of us, for it is impossible to traverse the infinite.

And if the world had no beginning and the number of men living[37] before, say, Socrates was infinite, and those living from Socrates until the present time have been added to it, there will be something greater than the infinite, which is impossible. And if[38] the number of men that have lived is infinite, the number of horses that have lived is certainly also infinite. [So] you will be doubling the infinite again. [And] if you add to these the number of dogs as well, you will triple the infinite, and if each of the other [species] is added it will be multiplied many times over. And this is among the greatest of impossibilities, for it is not possible to be greater than the infinite, not to mention many times greater. So if these absurdities, and, as we shall demonstrate elsewhere,[39] many others necessarily arise if the world is ungenerated, it is not possible that the world is ungenerated and without a beginning.[40] For in that case the infinite will be traversable and will exist in actuality, and the infinite will necessarily always be growing and never stop getting bigger than itself and be many times, not to say an infinite number of times, infinity.

So for the same reason then that he does not create more things God has not created the world from everlasting. And if it was impossible for the world to be created from everlasting, it is clear that no question remains as to why it did not exist earlier. For the further we 12,1
progress in our reasoning [about these issues], the more the beginninglessness of God's existence occupies our thoughts,[41] and as a result the same question nonetheless remains: why not earlier?

And if anyone insists on maintaining that the infinite should not be held responsible for more things not having been created, [even though] this has to be far and away the most scientific explanation,[42] but simply says that it is not only an infinite number of things that cannot have come to be but not even any more at all than [actually] exist, since nature only allows of as many creatures as have actually come into existence, in the first place he will not be able to give any reason for this. For if so many species have come to be, what would prevent another, say, two or three from existing in the world? For if men have managed to produce the mule by mating the donkey with the horse and have produced many mixed species of birds, how could it be other than absurd to claim than God could not bring natural animals of this kind into existence? (I call [these last] natural because men do not need to bring together different animals to breed them as in the other case,[43] but natural reproduction is preserved among them as among other [animals]). Given that such animals are in existence as a result of human techniques, why would it have been impossible for God himself to sow the seminal principles (*spermatikous logous*) [necessary] for their reproduction naturally? For in the case of the other [animals], as Aristotle says,[44] God, wishing to preserve the continuity of reproduction, 'filled up the whole in the only way that remained, by making generation continuous'. What, then, would have 13,1
prevented him from implanting the principle of reproduction naturally in such animals too – I mean in those that have arisen as a result of human husbandry?

However, even should we admit this, there is still nothing preventing us from using the same defense, [namely,] that the world has not existed from everlasting [unless] the infinite exists in actuality and there is something greater than the infinite, and [even] many times greater, and the infinite has become traversable; for these are impossibilities.

We have [now], then, also stated the reason why God, although he has always been good, has not created the world from everlasting.

4. Another argument: If God is good, and, as Plato in his wisdom says,[45] 'in the good no envy ever arises in regard to anything', and [it is] for this reason that he brought all things into being, why did he not bring any of [his] creations into existence in all respects identical to himself so that they shared in the same substance[46] and power with him? For either he could have, but did not wish to – and I shall not mention what would follow because it would be blasphemous (for he will not have escaped envy) – or it must be that, because he could not, he did not even wish to; but that would be to predicate weakness of God.

Actually, it would be impossible for any generated thing to be identical to the ungenerated cause that is responsible for its creation. If it were, God would have created himself and the uncreated would have turned out to be created – if, that is, [God's][47] substance is ungenerated by nature, but would nevertheless be generated for a second time if something that has been generated were of the same substance as he is.

So, if the impossibility of anything created being identical with its creator neither does away with [the doctrine] that God is good nor involves the consequence that he does not wish to make all things like himself nor inflicts any weakness or impotence upon him, then, when we hypothesise that the world is not everlasting, we neither do away with [the doctrine] that God is always good nor predicate weakness of his creative power. Rather we are hypothesising that the world cannot exist always because of the very nature of that which comes to be, both because it would have been impossible for the infinite to exist in actuality or to be traversable and because that which comes to be is not of a nature to be co-everlasting[48] with that which produces it. For if the world received its being and its substance from God through an act of creation (*dêmiourgikôs*), without in any way sharing in his substance, how could it have been coexistent with God from everlasting? For how could what always is be brought into existence? And if what is created is in all respects inferior to its creator – in substance, in power and in activity – it certainly must also be as a consequence inferior in its very existence. At any rate, we observe no [process of] generation, whether natural or artificial, in which the cause does not pre-exist[49] its effect. For the parallels they adduce for the world being co-everlasting with God, [by which] they show, as they suppose, that certain effects are coexistent with their causes, have nothing in common, as I see it, with the [case] under investigation. (The sun, they claim, being responsible for light, creates it just by being,[50] and the light is neither prior nor posterior to the sun nor the sun to the light;[51] and bodies in light are responsible for shadows that spring from them and are always coexistent with them). 15,1

5. In my opinion the second of these parallels should not even be worthy of consideration. For who is not aware that shadow is merely the privation of light? And privation in itself, as we showed a little earlier,[52] is non-being. So who with any sense would claim that body in light is the cause and instigation of what is in itself non-being? Non-being does not come to be; generation is of being. Just because privation of light is referred to by a word, being called shadow or darkness, one should not at once assume that it is also a kind of [real] existent or form. Deprivation of sight is also referred to by a word, being called blindness, and [a man or animal] that has maimed feet and is deprived of movement because of them is said to be lame, but nobody would say that blindness or lameness is a kind of form or [real] existent. So nor should one regard privation of light, even if it is called darkness or shadow, as a kind of form. What we were considering was whether there is any *entity* that has received its being and existence from something else through an act of creation and that is not posterior to that which brought it into existence. But shadow is the privation of light, not independent being. Therefore to say that body is the productive cause of shadow is the same as saying that death is that of non-being. So, just as we conceive of non-being not as a kind of existence, but only as privation of being, so too are we persuaded that shadow is merely privation of light and not a kind of existence imbued with form that is created by bodies. So how could one take the passing out of existence of light that arises from the interposition of bodies as a parallel to the generation and existence of 16,1 the world which stem from God?

And it is, moreover, quite clear to those versed in the mathematical sciences[53] that it is not even universally true that a shadow automatically accompanies a body in the light. For if the sun happens to be at the exact zenith, as is sometimes the case among those who dwell beneath the celestial equator and as far [north] as the Tropic of Cancer, it is clear that bodies under it which are in unbroken contact with the earth (so that there is none of the air in between in which shadow and darkness, like light, normally exist) then become shadowless. (Darkness, as Aristotle holds in the *On the Soul,*[54] exists in the same medium as light). Hence there needs to be time, during which [the sun's] rays do not reach under [things],[55] before [it] can, by passing the zenith, provide room at the sides [of things] for shadow. At any rate, even the earth, which always becomes shadowless under the impact of the rays [of the sun], little by little produces shadow as the sun's rays gradually withdraw.[56] And so, even if shadow were one of those things that exist in their own right, it would still be unable to serve as an image of the [phenomenon] under investigation. For it is not in every case [true] that a shadow coexists along with an illuminated body.

6. About the parallel drawn from the sun we have this to say.[57]

On the one hand, the substantial light that coexists with [the sun] in the sphere of the sun itself is formative and constitutive of the nature of the sun, as is clear to one and all.[58] But the essence and being of each thing is governed by its form. And nothing is its own creative cause. So nor is the sun the creator of its own light, since it is informed and has existence in virtue of it. And so, if someone should likewise say that the world is constitutive of the creator's being, let us concede that in that case the world stands in the same relation to God as the light within it does to the sun. But in that case God would not be the creator of the world, since neither is the sun [the creator] of the light which informs its being, if the sun has not in fact created itself. But if, on the other hand, all things that have been brought into existence by a creative act of the divine substance are extraneous to it and without a part in it,[59] the world does not stand in the same relation to the creator as the light in the sun does to the sun itself.

7. The light that is diffused in the air, on the other hand, is not always [one and] the same. It is agreed that it is perishable, for it is succeeded by darkness. And nor does it always remain the same, but, like movement and time, has its being in [a process of] coming to be.[60] Hence, as soon as the source of illumination[61] withdraws, the light in the air immediately passes out of existence (*eis to mê on*), just as there is neither movement nor time in the absence of a moving object. And so not even the very idea on account of which they used the parallel of light and the sun – I mean [the idea] that there exists something created by something else without its being posterior to that which produces it – is true. For if the same light does not continue [in existence], it clearly must be the case that the source of illumination always pre-exists the light produced in the air at any moment, for different light is continually flowing from the source of illumination at any given time like water from a spring. But God has, as Plato says,[62] brought into existence a single, unique world. And so, whatever light in the air actually is – for there is much debate about its [mode of] existence, whether it is an activity[63] of the source of illumination or a form or property of the air – since it is not always numerically one and the same but different at different times, the generation of light in air from the sun would constitute proof that there is [at least] one case of generation that is instantaneous and does not take place in the extension of time (in the way that nature, for instance, creates men over a certain period of time) and would [therefore] image the instantaneous creation of things by God, which occurred together with the thought;[64] but it would not also be an illustration of a created thing being co-everlasting with that which produces it because, whereas the source of illumination is one, the light is numerically different over time.

That the light in the air *is* perishable and numerically different over time the facts themselves confirm and Proclus teaches in his work *On Light*. Here is what he says:

> If light is either material or immaterial[65] in conformity with the difference between fire and the sun as sources of illumination, how, on the one hand, does the immaterial [kind] perish and how, on the other, does the material [kind] pass through material [objects]? For the whole of the air is clearly illuminated for us to the same degree whether [lit by light] from the sun or by earthly firelight; and it is certainly true that when a cloud passes in front of the sun the light is cut off and is completely absent on its other side.[66]

So [speaks] Proclus. And observe that he does not introduce the proposition that light perishes as a puzzle, but takes this for granted and derives from it a puzzle as to how, if light from the sun is immaterial, it perishes, and, when a cloud passes in front of the sun, is completely absent on the other side [of it].

And again, a little further on, the same [Proclus] says:

> How could the [light] in the heaven be continuous with that in the air? The one is perishable, the other not, and the one is connected to its source, while the other may be cut off from it and at times does not exist, and the perishable is not continuous with the imperishable. They are two [separate things] and differ in species.[67]

So even Proclus explicitly teaches that light in the air is perishable.[68]

And that light in the material of the heavenly spheres perishes in the same way as light in the air and is not, as Proclus claims, imperishable has, I believe, been clearly shown by the findings of astronomy. For if the cone of the earth's shadow comes into contact with part of the moon's sphere and passes over it (as eclipses, when the whole of the moon falls into the cone of the earth's shadow, palpably evidence), it is, one supposes, quite clear that the light diffused in the heavenly spheres is also perishable. For if, as the cone of shadow moves round, it spreads over different parts of the moon's sphere in turn and prevents the sun's light from reaching them (which is how occultations and eclipses of the moon occur), it is clear that light diffused there too is not, as Proclus claims, imperishable, but perishable. For if the whole heaven is of the same substance, and if the light diffused in the first and second sphere has been shown to be perishable, and moreover the light of the moon, which derives its existence from the sun, manifestly both comes to be and perishes, all at once during eclipses of the moon and little by little the rest of the time, it is surely clear that [dispersed light] will be perishable in the whole heaven as well. [And this will be so] even though, because the source of illumination is always present for the higher spheres since nothing casts a shadow over them, it seems, through its continuity, to be always one and the same light – as though a fire were continuously heating or illuminating its surrounds and the heat or light entering the air from the fire seemed, because of the continuity of the turnover, impervious to change.

8. And should someone claim that light generated in the air does not [ever] perish, but moves around with the source of illumination, first and foremost, such a notion is not true. Even when a cloud passes in front of the sun everything under it is thrown into shadow because the light previously generated in the air perishes even though the sun has not changed position. And so even when it does change position it does not carry [such] light around with it, but the light from the regions the sun leaves always perishes and new light is generated in the regions to which it moves.

And, more generally, even if one were to concede that the light in the air does travel around with the sun, it would clearly [have to] be inseparable from the source of illumination. And, if this were so, it would be a form or property of it. (And in fact Plato does hold that light is a form of the source of illumination.)[69] And if this is accepted, [their] parallel will be even wider of the mark.[70] For the sun, inasmuch as it is informed by it, naturally possesses its light inseparably;
21,1 for forms are inseparable from their substrata. But created things do not inform the substance of their creator. Therefore the sun is not the producer of light. For no substratum is the creative cause of the form that informs it, and nor for that matter is anything at all that possesses them the productive cause of its own property or attribute.

In Plotinian terms one would say the same thing [thus]:[71] 'light is an activity of the source of illumination going out to other things', for which reason he also says: 'when that which acts is present, its activity is present, when it departs, it departs along with it'. For if the light in air is an activity of the sun, either it is constitutive of its essence or it is not.

Now if, on the one hand, it is constitutive [of its essence], it necessarily always coexists with the sun; for the constitutive ingredients of essences always coexist with them. But the world is not a component of the essence of its creator. Therefore the world does not stand in the same relation to God as light does to the sun.

But if, on the other hand, the activity of light is not constitutive of the sun, it would be notionally possible for it[72] not to exist. But that is impossible. For if transparent objects, or, in general, those that are receptive of light, are present, it is impossible for them not to be illuminated in the presence of the sun; if they are not illuminated in its presence, it is no longer the sun.

Therefore if the light in the air is an activity of the sun, it must be constitutive for the sun.[73]

If, on the other hand, light is a property of the air, which to me at least seems closer to the truth than all other ideas about light, it would be an alteration to that which is transparent, or, in general, to that which is receptive [of illumination]. And Aristotle in fact says[74] as much in the second book of *On the Soul*.[75] The passage goes as follows:

> Light is the activity of this – of the transparent *qua* transparent. And potentially, where it is, there may also be darkness. Light is a kind of colouration of the transparent when it is actually transparent through the agency of fire or something similar, such as the body above;[76] for this is one and the same in these.[77] So what the transparent is and what light is has been stated.

This then is what Aristotle has to say about light. And he is clearly stating that light is a kind of form or perfection of the transparent. So, if the transparent is in air or water or the like, it is clear that light is a transformative property of such bodies. This being so, it is not surprising that, if the one is suited to acting and the other to being acted upon, as soon as the agent and the patient come together, the one acts and the other is acted upon. After all, fire too, when in the proximity of air, which is [always] potentially hot, alters it and makes it actually hot. And if, when the source of illumination moves away, the light that is produced in the air also disappears, there is nothing strange about that; for it is also the case that when fire moves away heat ceases to be produced in the air, the only difference being that in the case of light its passing out of existence, like its generation, is swift, or even instantaneous. So if the sun alters the transparent [only] qualitatively, but God is productive of the substances themselves, and if alteration is different from generation and the creation of substance, either let them show in the case of generation in the strict sense (i.e. that of substances) that that which is caused is coexistent with its cause, or let them concede that even God is not a creator of things and does not bring their essences into existence but only causes alteration in things which are co-everlasting [with himself].

And besides, they will not [be able to] show that the cause is co-everlasting with that which is caused even on that basis. If different effects are produced at different times in air and the [other] things that are receptive of light, and the light produced in them – if, as has been demonstrated and as Proclus believes, it is perishable – is not [always] numerically one and the same, then it is not the case that light, remaining one and the same, coexists with the sun, any more than that heat in the air originating from [a fire], remaining one and the same, does with the fire – if, that is, the sun is one and the fire is one while the light in the air, and likewise the heat, is different at different times.

After the above demonstrations it is clear that light and the sun cannot provide an illustration of something created being able to coexist from everlasting along with its creative cause. So even the plausibility afforded their false [hypothesis] by such an illustration is lost to them. So either let them show that there is something among created things which as soon as its creator exists also exists along with him, or, if all created things, whether man-made or natural, are later[78] in time than what produces them, then it is necessary on that basis too that the world not be co-everlasting with God.

The End of the Refutation of the First Argument

24,1 The Second Argument of Proclus the Successor[79]

The second [argument]: If the pattern of the world is eternal, and if its essence is being a pattern[80] and it has this power[81] not accidentally but in itself, being a pattern by its very being,[82] then, since it is eternal in its being, it would presumably be eternally a pattern. And if being a pattern is present to it eternally, there would necessarily always be a copy too; for a pattern is relative to [its] copy. But if the copy [ever] did not exist, when it did not exist the pattern too did not exist, and it will not exist [at any time in the future] when the copy does not exist, unless, there being no copy, it is not a pattern at all, or it is not the pattern for the copy [in question];[83] for neither of [a pair] of things described as correlative exists if the other does not. If, then, the pattern is eternally a pattern, the world always is, being the copy of an eternally existing pattern.[84]

The Sections of the Refutation of the Second Argument

1. That the second proof[85] of Proclus is constructed from two hypotheses, the one, which states that the world has come to be in relation to an eternal pattern, being Platonic, the other, which Proclus puts forward as Platonic although it is not Plato's, [stating] that the pattern of the world has its being in being[86] a pattern.[87]

2. That there is manifestly disagreement between Aristotle and
25,1 Plato over the hypothesis of Forms, Plato holding that there are Forms of things and Aristotle seeking to demolish the hypothesis.

3. That, even if one concedes that Plato's hypothesis of Forms is true, it is not also true, as Proclus thinks, that they have their being in being patterns, since Plato refers to them as substances.

4. That, if it is not the essence of the Form of the world, in relation to which [the world] has come to be, to be a pattern, no further necessity remains, even if the pattern in relation to which the world has come to be is eternal, for the world too to exist always.

5. That, even if one were to suppose that the Forms are not substances but the creative principles in accord with which things have come to be, not even then is there any need for created things to coexist together with the creative principles; on the contrary, the necessity is rather that the principles of things should pre-exist them. Including [a demonstration] that God's creative activity and knowledge extend even to the most individual [of things].

The Refutation of the Second Argument

1. Plato hypothesised that there are Forms or patterns of existing things, looking to which the creator makes the world and everything in it as though [producing] copies from certain archetypes, and certainly states that this pattern of the world is eternal.[88] Proclus, using this axiom of Plato and himself adding a second hypothesis, [namely], 26,1
that the Forms have their essence in being patterns – something which Plato clearly nowhere states – concludes from these two hypotheses that it is necessarily the case that the world always exists; for copy and pattern are relatives, and relatives are destroyed together or exist together.[89] So if, he says, the pattern of the world is eternal, and [if] its essence and substance consist in its being a pattern and it does not have its being in one way and its being a pattern in another, and [if] a pattern is a pattern of a copy, there is every necessity that the copy too of which it is the pattern, that is, the world, should always exist; for if the copy does not exist for ever, neither will the pattern be eternal. So if Plato has clearly stated the one [hypothesis] – that the pattern of the world is eternal – he is of necessity conceding the other as well – that the world, the copy of the eternal pattern, always exists. Such is the second of Proclus' proofs.

2. But perhaps we should have taken more account of the facts and not simply of Plato's hypotheses and not have come to grips with this argument until the philosophers had resolved their differences with each other on the subject of the Forms; for who does not know how great was the conflict and disagreement between the [two] foremost philosophers on this very subject?[90] For throughout his writings, whenever the occasion arises, Aristotle, Plato's pupil, criticises his teacher's hypothesis of the Forms in the strongest terms, and 27,1
fills book Alpha of the *Metaphysics* in particular with criticism of his views in regard to them. In it, giving an account of Plato's ontology,

he says that Plato fell into his notion of the Forms through a misunderstanding of the teaching of Socrates. The passage goes as follows:

> After the [philosophies we have] named came the work of Plato, which in most respects followed these [thinkers], but also had peculiarities [which set it apart] from the philosophy of the Italians. For, having been a pupil of Cratylus from an early age and [become familiar with] the Heraclitean doctrines that all perceptible things are always in flux and that there is no knowledge of them, he held these views even in later years. Socrates, on the other hand, had concerned himself with ethical matters and not at all with any aspect of the world of nature, yet was seeking the universal in them and was the first to have paid attention to definitions, and after taking him as a teacher, [Plato], on account of such [considerations],[91] took the view that this[92] had to do with other things and not with anything perceptible;[93] for a common definition cannot be of any perceptible thing, since they are always changing. Accordingly, he called such things Forms and [held that] all perceptible things are separate from them and named after them; for, [he held], the many things that have the same name as the Forms exist by
> 28,1 participation [in them].[94]

In this passage we hear Aristotle state clearly that Socrates was the first to pay attention to the methodology of definition and that he sought out and defined the common features in perceptible things. [He would ask], for example, what knowledge as such is as seen with common features in each of the particular branches of knowledge, in astronomy, for instance, and geometry and the rest, and what beauty as such is, what justice is, and similarly with everything. [But, he continues,] Plato, taking up this method of Socrates, assumed that Socrates was talking about things other than and separate from perceptible things and not about perceptible things themselves and the common features in them; for, he says, [Plato], as a holder of Heraclitean opinions, believed that there could be no definition of perceptible things, since there could not even be knowledge of them because they all have their being in a state of flux and change and no perceptible thing [ever] has stability or definition. (It is because of this that he says in the *Timaeus*[95] that all perceptible things come to be[96] and perish, and that, never truly being, they can [only] be grasped by belief along with irrational sensation and not by knowledge). Believing, then, for this reason that Socrates was creating definitions of common features associated with certain separate entities [that existed] apart from perceptible things, he called[97] these Forms and patterns of perceptible things, [holding that] perceptible things exist by participation in them and are called by the same names as they are. For both the Form of animal and the perceptible animal itself are called animal and the same applies to man and to the beautiful and the just, etc.

From this we can most certainly see that Aristotle's refutations of Plato are not directed at people who have misunderstood Plato, which is a fiction created by some more recent [commentators] out of embarrassment at the disagreement between the [two] philosophers,[98] but [rather] constitute a rebuttal of the notions of Plato himself. For, if Aristotle had not been attacking Plato's own position on the Forms but, as these [commentators] claim, [that of] people who had misunderstood[99] him, he would have specified precisely this at the outset and not have refuted the doctrine of the Forms generally and without qualification, just as, for instance, in the passage we have [just] quoted, he does not criticise Socrates, who defines the common features in perceptible [things], but Plato, who has not understood Socrates correctly.

You will find that this is the teaching method he practices in all his own writings. In the physical treatises, for instance, when criticising the views of the physical thinkers who came before him on the principles from which things are composed, he did not demolish [the position] that there are principles and elements of bodies but refuted the hypotheses of early thinkers about the principles and then instructed us in his own ideas about them. And when, in *On the Soul*, he was criticising [thinkers] with false conceptions about the soul, he did not totally demolish [the position] that soul exists but [only] that it is as these people believed.[100] [And] similarly in his writings on growth and generation and place and time and everything else he first demolishes false views on the subject at hand and then sets out what he believes are the correct ones.

He should not then in his account of the Forms, if he accepted that they are as Plato believes, have attempted to demolish the hypothesis totally but should [instead] have explained how some people had misunderstood Plato's teaching and what the correct understanding of them [sc. the Forms] is. This he has clearly nowhere done. On the contrary, in addition to [levelling] criticisms [at them], he frequently scoffs at the notion [of their existence]. In the second of his [works] on demonstration he says: 'For we can say goodbye to the Forms; for they are [mere] twittering',[101] that is to say, mere vocal noises that sound like words but are devoid of any reference to things. And, again, in [book] Alpha of the *Metaphysics* he declares that statements about them are empty talk and poetic metaphor.[102] And, moreover, when about to launch once more into criticism of them in the introductory part of the *Ethics*, he first apologises out of respect for his teacher, saying that inquiry into these matters is arduous because he is forced to rebut the views of a man who is dear to him – he means Plato. But, he says, although Plato is dear to him, it is pious to value the truth more highly,[103] and, indeed, it is right, he says, for us as philosophers not only to refute the hypotheses of others in the interests of preserving the truth, but also for each of us [to refute] our own if they are found not to be in agreement with the truth.[104] The passage goes as follows:

> Perhaps we had better consider the universal [good] and look into the puzzles about what is meant by it, even though the inquiry becomes arduous because it is friends of ours who have introduced the Forms. [For] it would perhaps be thought the better course, even our duty, to demolish even our own [positions] in the interest of preserving the truth, especially as we are philosophers; for, while both are dear, it is pious to honour truth above [our friends].[105]

And even Proclus himself has admitted, showing respect for the truth, that the [two] philosophers are in disagreement on many points and most especially in regard[106] to the hypothesis of the Forms. At all events, in the work he has entitled *An Examination of Aristotle's Criticisms of Plato's Timaeus*,[107] in the first section, he writes, to quote his exact words,[108] as follows:

> Aristotle has difficulties even with the very word 'pattern', saying that it is metaphorical,[109] and is much more opposed to the doctrine that introduces the Forms in general and especially to the living-creature-itself, as he has written in the *Metaphysics*.[110] It is probable that the man did not reject anything else of Plato's as [categorically] as the hypothesis of the Forms. Not only does he call the Forms [mere] twittering in the logical works,[111] but he attacks the Good Itself in the *Ethics*,[112] does not think it appropriate to refer cases of generation to the Forms in the physical works, as he states in *On Generation and Corruption*,[113] [and rejects them even] more comprehensively in the *Metaphysics*, in that, in the course of discussing the first principles, he spins out lengthy denunciations of the Forms at the beginning, in the middle and at the end of that treatise, and in the dialogues, where he states explicitly and vehemently that he cannot sympathise with this doctrine even if people are going to think that he is criticising it out of contentiousness.

Thus even Proclus himself has explicitly conceded the disagreement between the [two] philosophers, or rather, demonstrated it from Aristotle's own [writings]. This being so, one might well be amazed at the gross effrontery of those who have tried to show that Aristotle and Plato are in agreement even on this point. So how could it be other than superfluous for us to reply to a proof derived from an hypothesis which is the subject of such great disagreement even among the very foremost philosophers? For their successors too have consistently followed the founders of their respective schools (*tês oikeias haireseôs*).[114] However, so as not to seem to be avoiding [the task of producing] refutations [of this argument], especially when we have promised[115] to support Plato on the issue of the generation of the world, let us concede his position on the Forms and see, for the rest, whether Proclus is correct in his assumption that it is the essence of the Forms to be patterns. For if it can be exposed as a false assumption, Proclus will no longer be able to conclude from the hypothesis of the Forms that the world has existed from eternity alongside its own pattern[116] or that, because Plato assumed the existence of the Forms, he also thought that the world was everlasting.

Let us [then] begin our arguments of refutation with the following.

3. It is necessary that the Forms of perceptible things, or [their] patterns, should either be substances or not be substances.

Now if they are substances, as Plato believes – for he not only calls them substances, but 'first substances';[117] for he calls the pattern of the world 'living creature itself' and 'perfect living creature' and [describes it as] 'truly existent' and 'always being the same and unchanging';[118] and besides, if the copy of the pattern is self-subsistent and a substance, then all the more would the archetype itself be a substance and self-subsistent – if, then, the Forms are substances, it is clear that they do not have their being in [their] relation to something else; for a substance is something self-subsistent and in need of nothing else for its existence.[119] So if perceptible substances, [which are] their copies, exist in their own right, not having their being in [their] relation to something else, least of all will those [substances] that are primary and always self-subsistent and the same have their being in [their] relation to something else. So if the pattern of the world is a substance, the consequence is[120] that it would have its existence in its own right and not in relation to something else. And if it does not have its being in relation to something else, then it is not its essence to be a pattern; for, as Proclus himself says, a pattern is a relative, for it relates to a copy, and no substance is a relative.

Further, if the Forms are relatives, as Proclus believes, and it is their essence to be in a relation to perceptible things, whose patterns they are, as right is related to left and master to slave, and if relatives are not substances (for relatives *qua* relatives are accidents), then the Forms are not even substances, but accidents. But they were substances *ex hypothesi*.[121] Therefore they are not relatives. But a pattern is a relative. Therefore it is not the essence of the pattern of the world to be a pattern and to have its being in relation to its copy, the world.

Further, if the Forms are relatives, and if relatives do not have their own existence, but have their being in other [things], as was shown in the *Categories*[122] – for they are either in substance, as [in the case of] father and son, [or] master and slave (for the relation between these occurs in the substance man), or in quantity, as [in the cases of] many and few, greater and lesser [and] double and half, or in quality, as [in the cases of] whiter and sweeter and wiser, or in one of the other [categories] – it is, one assumes, necessary for the Forms, if they are indeed relatives, to exist in some other subject. And the same goes for the world, for it too is a relative. For, just as the pattern of the world has its existence in being a pattern, so, one assumes, must the copy of the pattern, the world, have its being by virtue of being a copy. But an image is a relative. Therefore the world will be a relative in virtue of its very being. And if this is so, [both] the Forms and the world must be accidents and not substances. But the former are substances *ex hypothesi* and the world too is a substance. Therefore they are not relatives.

Further, relatives are cases of opposites.[123] But no substance is a case of an opposite. Therefore substances are not relatives.[124] But Forms are substances. Therefore Forms are not relatives. But a pattern is, as stated,[125] a relative, for a pattern is [related] to a copy. Therefore Forms do not have their being in being patterns. Therefore it must be, if they are indeed substances, that their being belongs to them in one way and their being patterns in another.

4. But if the Forms do *not* have their being and existence in being patterns, there is no necessity that, because the Form in relation to which the world came to be is eternal, the world too should always exist. Plato says[126] that the heaven or world is the pattern for civic well-being and that the true statesman looks to the order in it when arranging the affairs of the state. But, since the heaven does not have its essence in being a pattern for civic well-being, there is no necessity that, because order exists in the heaven, its copy, I mean a successful city, should at once exist. [And] in just the same way, one assumes, even though the Forms are eternal, it is not at all necessary that their copies (I mean the perceptible substances) should also always exist,[127] since their being does not consist in being patterns if they are not relatives but substances. The necessity in these matters holds in the other direction. If there is a copy, a pattern must exist; but, even if a pattern exists, though not [merely] as a pattern but because it is also a substance, there is no necessity for a copy to exist as well. The king himself is the subject of a royal portrait, but this does not mean that as soon as the king exists a portrait of him must also exist.[128] It is one thing for the king *qua* king to be a man, another for him to be the subject of a portrait. Whenever he *is* a subject, then in every case there is also a portrait, just as whenever he becomes a father, a son is always implied as well. But there is not immediately the subject of a portrait as soon as a king exists, just as someone is not immediately a father and a master, or on the right hand side, or anything else[129] that falls under [the category of] relatives, as soon as he is a man.

5. These then [are the consequences that follow] if they assume that the Forms are substances, as Plato believes they are.

But if they will claim that they are not substances but certain creative principles or concepts in accord with which the creator frames things – for what else could they be if they are not substances? – even so there is no obvious necessity that products based on these creative principles should in every case coexist together with them. Once more the necessity is in the other direction. If [such] products exist, the principles in accordance with which they came into existence must in every case exist as well, but [created] things will not in every case follow upon [the existence of] the principles. A shipwright or a builder may be in possession of the principles for [building] a ship or a house but not yet be creating [anything] based on them. The being of such principles consists in their being concepts of a certain kind, but when the creator acts in accord with them, they become patterns as a consequence [of that]. If, then, products based on them do not in every case immediately follow upon [the existence of] creative principles, nothing prevents the world from not always existing even if the creative principles for the world are eternal.

And that it is not necessary that the things of which the principles are principles should coexist with them, but rather, on the contrary, that the principles should pre-exist the things, one may also learn from the following.

All would agree that God has accurate foreknowledge of the future. Among the Hellenes[130] indeed shrines and oracles[131] were established in every corner of the land [where] an evil demon feigned the true foreknowledge of God.[132] And it is moreover a shared conception of all who in any way recognise the goodness and ineffable power of God that all things benefit from the providence that flows from God and not even insignificant things (*to tukhon*) escape his attention. And in his work *On Ten Puzzles About Providence* Proclus is clearly in complete agreement with us on these matters, for in it he says this:

> Since, as has been stated, providence is determined in accordance with the One and the Good, and the Good is prior to Mind (for Mind desires the Good, as, indeed, do all things, and not the Good Mind), it is necessary that providential knowledge too should be above intellectual and that providence should therefore know all things by its own unity, through which it also makes all things good, those that know and those that do not know, those that have life and those that do not have life, those that have being and those that do not have being, projecting unity upon all through the reflection of its own unity.[133]

And a little further on:

> In summary, then, we claim that this unity produces all things and sustains all things, and that it has an existence more real than all being and clearer than all knowledge, not being dispersed by the objects of its knowledge nor moving among them. It is the knowledge of the soul and of the mind that show the characteristics of these last [conditions], for every single mind is many in its [mode of] being and in its knowing, and every soul, being movement, also knows by means of movement. But it [sc. the unity of providence], remaining stationary and undivided in [its] unity, also knows all things in this same mode, and not only man and sun and everything whatsoever of that kind, but also each particular thing. For nothing escapes that unity, whether it is a question of its being or of its being known.[134]

And a little further on:

> Thus the unitary knowledge of providence is, in the same undivided [entity], the knowledge of all divided things, both of all that are most individual and of all that are most universal; and just as it has caused each thing to exist in unity, so does it know each in unity.[135]

We have, then, heard Proclus clearly stating that the first cause of all things, which is the object of desire for all things – that is, the Good and the One, for they characterise the first [cause] in this way alone, if one should affirm anything at all of it – both brings into existence and has knowledge of even the most individual of things and extends its providence to all things.[136]

And the philosopher Plotinus in *On the Intelligibles, or, On the Good*[137] says similar things about the creative and providential power of God in these words:

> Nevertheless, if it is necessary that each activity [of his] should not be imperfect and it is not right to think of anything belonging to God as other than whole and entire, all things must be present in anything of his. So existing for ever must be too. So the future must already be present too. There is no 'later' in that [place]; what is already present there comes to be later elsewhere. If, therefore, the future is already present, it is necessarily present as having been provided for with a view to later [time]; that means that nothing will need for anything then, that nothing will be left out. All things, then, already existed and always existed and existed in such a way that one could say later that this exists on account of[138] that. When it is

> extended and, as it were, unfolded, it can reveal this after that, but when it is together, it is all 'this';[139] that is, it has its cause too within itself.

Plotinus too, then, states that not only things that are, but things that will be are already present to God's foreknowledge; for he states that what has existence later among generated [things] is always present in him. Then he adds the manner in which the future is present to God: that it is by thinking even what is not yet in existence. 'It is necessarily present', he says, 'as having been provided for with a view to later [time].' So if even all things that do not yet exist are present as existing things to God by virtue of his foreknowledge, he doubtless knows in what manner and according to which principles each future thing will come to be, and once it comes to be, for what reasons it will undergo the consequences that flow from unerring providence in each choice of a life [it makes]. 'For all things', says Plotinus, 'already existed and always existed.' For the 'already' shows that things that will exist later are present for God before they come to be. And he adds 'and always existed' so that nobody will think that 'already' means there is a beginning. 'And existed', he says, 'in such a way that one could say later that this exists on account of[140] that.' [He is referring to] things the causes of whose generation we know after their coming into existence, as for example that man alone of animals here [on earth] has the use of hands on account of the activity of reason, since the creator has made hands as tools for the reason. So what we know [only] after [its] coming into existence – [i.e.] that this exists on account of that – has always had prior existence in God's knowledge.[141] And the same applies not only to whole species, but to individuals with separate existence as well; at any rate, explaining himself, he adds 'when it is extended and, as it were, unfolded, it can reveal this after that', [by which he means] when it is brought to generation, because, to take an example, after the mixing of the elements, comes condensation and alteration, and after that shaping, and after that cooling, and similarly with all things. 'But when it is together', he says, 'it is all[142] "this", that is, it has its cause too within itself.' By 'together' he means in God's foreknowledge; for in it the principles of all things are present [as a whole] without division.

So much, then, is demonstrated concerning the prescient and provident power of God out of [our] common conceptions and the [works of the] philosophers we have cited.

And, in addition to this, the belief in regard to fate,[143] whatever it may be – for there are various doctrines in regard to it – is believed by the Hellenes[144] to grasp all of the causes of future happenings together under one time, that of the date of birth,[145] and foretell them, or even bring them about.[146]

So, given that we are accepting these [positions], on the basis of either doctrine – I mean the one about providence or the one about fate – there is every necessity that the principles of future happenings should both be foreknown and pre-exist [those happenings]. And, this being so, it is clear to all that there is no necessity that the things [themselves] should coexist with the creative principles and causes of created things, but that it is, on the contrary, necessary from what we have shown that their principles should pre-exist all created things. And so, even if the Forms and patterns of things are certain ideas and principles of the creator, in accordance with which he has brought the world into existence, it is certainly not necessary that the world itself should coexist from everlasting along with God's knowledge about the world.

And nor, then, will any argument compel Plato to hypothesise that the world too has always existed because he says that the pattern of the world is eternal. For it has been shown that the being of Forms does not consist in their being patterns.

The End of the Refutation of the Second Argument

42,1 **The Third Argument of Proclus the Successor**

The third [argument]: If the creator is the creator of something,[147] either he will always be an actual creator, or sometimes [only] a potential one [and] not always be creating.

Now if the creator is always an actual creator, that which is created will likewise always be actually undergoing creation. For when, says[148] Aristotle, a cause is actual, that which is caused will likewise be actual; if that which builds is, so will that which is being built, [and] if that which creates health is, so will that which is being made healthy. And Plato [says][149] in the *Philebus* that a producer (*to poioun*) is a producer of something which *is* coming to be, not that a producer [is a producer] of something which is *capable* of coming to be, or that that which is *capable* of producing [produces] something that *is* coming to be.

But if, on the other hand, that which is created is not actual, nor will that which creates be actual. And if it is not actual, it will be potential, being capable of creating before it creates. And everything that is potentially something, says[150] the same [Aristotle], becomes actual from something that is actually what it is potentially, the potentially hot from the actually hot [for example], and [similarly with] the cold, the white and the black. Therefore the creator will be an actual creator after having previously been a potential [creator],
43,1 there being another actual creator that makes this previously potential creator an actual one. And if the former is always the actual cause of the latter's being a creator, the latter too will always be a creator on account of the first[151] axiom, the one that states that whenever a

cause is actual that which is caused is also actual. And therefore there is always something undergoing creation. But if the former too was at some time [only] potentially responsible for making the creator create, it in turn will need some other [cause] to make it actually make the creator create on account of the second axiom,[152] which states that everything potential has need of the actual if it is to become actual. And the same argument will apply in turn to this one, and we shall either go back indefinitely seeking one cause before another for each successive potential cause becoming actual,[153] or we shall end by agreeing that there is always some cause that is actual.

But once this is agreed, it follows that [its] effects are always actual too and that the world is always undergoing creation, if indeed [it] also [follows] that the creator is always a creator, this having been demonstrated[154] with the aid of two axioms, the first that one of a pair of relatives will always be in the same condition as the other, potential if it is potential, actual if it is actual, the other that everything that exists potentially changes to[155] actual [existence] through the agency of something that is actually what it was previously potentially and subsequently actually.

The Sections of the Refutation of the Third Argument 44,1

1. That in this section Proclus has cobbled together a fallacious argument which depends on the ambiguity of potentiality and actuality.

2. A setting out of the different senses of potentiality and actuality.

3. That an actual effect does not in every case necessarily attend a cause 'in capacity'[156] (that is to say, [actual] in the first of the senses of actual). This being so, it is possible for God to always be an actual creator in capacity but for there not always to be a world.

4. That something[157] advancing from second potentiality, the kind that is based on capacity [and] which is also first actuality, to second actuality does not need anything else to bring it from [mere possession of] the capacity into actuality.

5. A setting out of Proclus' words and a demonstration from them that he has cobbled together a fallacious argument by using the different senses of potentiality and actuality without qualification. Including [a demonstration] that God, by being an actual creator in capacity, neither necessarily implies[158] a world that coexists with him from everlasting nor will have need of some other creator[159] to bring him from capacity to actuality.

6. That not even in the case of first potentiality is it universally true that the[160] potential in every case needs something else which is [already] actually that which it[161] is [only] potentially in order to be brought into actuality, but only something that is actual, but not in every case of the same kind.

45,1 **7.** That it is possible for nature to be brought even of its own resources from fitness[162] for something to a capacity and actuality in relation to it and not to need any external agency to accomplish this.

The Refutation of the Third Argument

1. That the philosopher's proof is sophistical. What he *should* have done was respected the dignity of philosophy and employed the utmost dialectical precision. Since he was investigating matters of such importance, he [should] not have misled us with ambiguities as sophists do the young. Rather, with a care for the truth, knowing[163] that potentiality and actuality are ambiguous words, [he should] have distinguished their different senses, and explained to which of the senses of potentiality the first of the axioms proposed by him conforms, and then to which of those of actuality[164] the second [conforms], and next, hypothesising each of the senses of potentiality and of actuality[165] in relation to the creator, have investigated the consequences of each hypothesis.

Let us, then, making our only goal the discovery of the truth and following the rules of dialectic, first distinguish the various senses of potentiality and actuality, and, having done so, consider what truth there is in the axioms of the philosopher based upon them from which he has constructed [this] the third of his proofs.

In addition, for the sake of those who are inexperienced in this kind
46,1 of argument it will be a good idea to provide a brief explanation of the concepts of potentiality and actuality.

2. One should be aware, then, that Aristotle quite explicitly taught[166] that potentiality is twofold and that actuality is twofold, and this is accepted by everyone. In one way it is that which has a kind of natural fitness to [become] something that is said to be potentially that thing, as we say that a child is potentially a grammarian[167] because its nature is receptive to acquiring from others[168] the theorems[169] of grammar at the right time. This is one of the senses in the case of potentiality. It is called first potentiality, [the kind] that is based on fitness. And when someone has come into possession[170] of the theorems of grammar and has the knowledge of them in his mind, but is not active with regard to them or externalising[171] his reflection[172] on them, perhaps because he is eating or because his thoughts are engaged elsewhere or because he does not want to, he too is said to be potentially a grammarian, but not in the same way as the first person, for that person had not come into possession of the theorems, but was said to be potentially a grammarian only through being naturally disposed to acquire them.[173] This sense [of potentiality] is called second potentiality, or potentiality based on capacity.

And the very same thing that is called second potentiality, or potentiality based on capacity, is also called first actuality. For, in contrast to first potentiality, the [kind] based on natural fitness, whose possessor was said to be potentially a grammarian by virtue of being naturally disposed to acquire the theorems [of grammar], a person who has already come into possession of the theorems would already be an actual grammarian, having already acquired *actual* possession [of them].[174] But since there is also a second kind of actuality, under which a grammarian is currently (*êdê*) presenting the theorems, whether in passing them on to someone else or practicing on his own, by contrast with this current (*êdê*) actuality, a person who possesses the theorems only in capacity is, as I have stated,[175] said to be a potential grammarian – potential in the second sense, [that is]. So in this way the same person – one who only has possession of the theorems of grammar but is not making active use of it[176] – is described both as a potential grammarian and as an actual grammarian, but potential in contrast with second actuality, under which he presents the theorems when teaching, and actual in contrast to first potentiality,[177] under which he did not yet have possession of the theorems, but was said to be a potential grammarian because he was capable of acquiring it at some time in the future.

3. These things being so, and potentiality being twofold, and actuality likewise being twofold, and second potentiality coinciding with first actuality, the first of the axioms, the one that says 'when a cause is actual, that which is caused will likewise be actual',[178] is true only for second actuality, not for first, which is also called second potentiality. For when a teacher is currently teaching, there must also be someone being taught, and if a builder is currently building there must also be something being built. But the axiom is not[179] true in the case of first actuality. For it is possible for a builder to have the capacity to build but not be building and for someone who has the capacity to teach not to be teaching. And so, whereas a cause in this sense, I mean a builder or teacher by capacity, is actual, an effect, that is, something being built or taught, is not so in every case.

I could have gone on to cite the foremost philosophers, including, among others, the Aristotelian commentator Alexander in his commentary on the *Apodeictics*,[180] saying almost word for word just what I have. But, since the matter is clear, I shall refrain from such a course at the present time.

This, then, is the truth in regard to the first of the [two] axioms.

4. The second [axiom], on the other hand, the one that says that when a potentiality is brought to actuality it must in every case change and be brought to actuality through the agency of something else which is actually what it is potentially, only applies to first potentiality. For the child who is a potential grammarian changes to being an actual grammarian through the agency of someone who is

an actual grammarian, and water that is potentially hot changes and becomes actually hot through the agency of, say, fire or, [more] generally, something that is actually hot. But this axiom is not true in the case of second potentiality. For someone who is already in possession of the theorems of grammar, and who is said to be a
49,1 potential grammarian in that sense, does not, when he changes from this [kind of] potentiality to what is known as second actuality, in which [state] he is currently presenting the theorems [of grammar], change through the agency of another who is already an actual grammarian, but changes from being inactive to being active by himself, without any intermediary. And the same applies in all other cases.

Now that we have elucidated these matters, let us get on with the matter in hand, starting by quoting the philosopher's [actual] words:

5. 'If the creator', he says,

> is always an actual creator, that which is created will likewise always be actually undergoing creation. For when, says Aristotle, a cause is actual, that which is caused will likewise be actual.[181]

Observe how he does not specify with reference to which sense of actuality he is hypothesising that the creator is an actual creator, but simply and without qualification says that an actual effect follows upon an actual cause, so that he can predicate of every case of actuality an axiom that is [in reality only] true of one sense of[182] actuality, which we have shown to be a fallacy in that it has application only to second actuality. If, then, we hypothesise that God is always an actual creator in capacity, that is, first actuality, what further necessity is there, if God is always an actual creator in this sense, that the world too should always exist?

The philosopher has acted thus with sophistical intent so that,
50,1 when we say that the world does not always exist, he may at once totally eliminate [the possibility] that God is at the same time (*kai*) always a creator, and [so that], once it is agreed that God is in no sense an actual creator before he creates, he may [quite] consistently infer that he is a potential creator in the first sense of potential. At all events, hear what he goes on to say:

'If, on the other hand', he says,

> that which is created is not actual, neither will that which creates be actual.[183]

Observe again how he uses 'actual' without qualification and does not add with application to which sense of 'actual' the[184] axiom is true. [In fact,] it is clear that it is true for second actuality but not for first.

Even when the house does not actually exist nothing prevents the builder from actually existing in capacity. For a man who has the capacity is a builder even when he is not building, and likewise a doctor even when he is not practicing, but it is not possible to actually build or practice medicine if there is not also in every case something being built or someone being treated.

And it is clear from what he goes on to say that it really was from a wish to mislead by means of [this] ambiguity that he did not distinguish the different senses of potentiality and actuality, so that, as I have said,[185] when we say that the world does not always exist, he can infer that [in that case] God was not in any sense an actual creator before he created the world but a potential one in the first [sense of potential]:

'And if it is not actual', he says,

> it will be potential, being capable of creating before it acts. And, 51,1
> according to the same writer [sc. Aristotle], everything that is potentially something becomes actual from something that is actually what it is potentially, the potentially hot from the actually hot [for example], and [similarly with] the cold, the white and the black. Therefore the creator will be an actual creator after having previously been a potential [creator], there being another actual creator that makes this previously potential creator an actual one.[186]

Observe how, as we have stated,[187] after eliminating [the possibility] that God was an actual creator on the ground that the world did not actually exist, he immediately infers that he was a potential creator in the first [sense of] potential, as though every sense of[188] actual has been eliminated. At any rate, he claims that there will be need of another actual creator to make this previously potential one into an actual creator – a thing that we have shown does not apply to something that has a capacity, but to something that only has fitness. And yet, if the world does not actually exist, it is not immediately implied that the creator is a potential creator in the first [sense of] potential. As I have said,[189] capacity [on his part], that is to say, first actuality, could equally be implied. But the philosopher sophistically claims that what is said of first potentiality (I mean that it has need of something that is [already] actually what it is [only] potentially to be brought to actuality) follows for potentiality generally and without qualification, and then, for the rest [arguing quite] consistently, adds a lot of rubbish, stringing together creator after creator as he draws 52,1
the conclusions which follow from his first absurd hypothesis.[190]

But, since nobody with any sense at all would claim that God is a potential creator in the first [sense of] potentiality before he creates, once this absurd premiss has been eliminated, this whole tissue of spiders' webs, the philosopher's syllogisms, collapses. For if God is a creator in capacity before he creates, he will neither have need of another to bring him into actuality, nor, since it is in *that* sense[191] that he is an actual creator, does he on that account straightaway bring with him a world that coexists with him from everlasting. Just because there is someone with the capacity to build,[192] there will not in every case also be a house. Nor, when a builder passes from [mere possession of] the capacity to its exercise, will he have need of another, actual, builder to bring him to this.

Aristotle himself has adequately shown[193] in the *On Interpretation* that to use ambiguities to create fallacious arguments is sophistical and most offensive to anyone with any sense, describing this kind of fallacy especially as an annoyance:

'I call "opposing" ', he writes,

> [a statement which contradicts another by affirming or denying] the same thing of the same thing – not ambiguously, and with such other [restrictions] as we specify to counter sophistical annoyances.

And it surely is the case that arguments which, neglecting the facts, strive to refute those they are arguing against by purely verbal means[194] are in truth an unwanted annoyance – something the great Proclus has not shrunk from doing to us even in these [important] matters.

53,1 **6.** It is worth observing how in his very selection of an axiom the philosopher establishes for himself an adequate supply of materials for his fallacious arguments. For, so that he can string together his great nonsense about creators – of a potential creator that has need of another, actual, creator that will make it too an actual creator, and of this [second] creator in its turn, because the earlier one was potential,[195] being potential itself and having need of a third, actual, creator, and of this one in turn [having need] of another – and so that he can in this way continue this absurdity *ad infinitum*, having said that everything potential has need of something else that is actual to bring it to actuality, he says that this actual thing must be of the same kind as *it* is potentially; for example, if something is potentially hot, that which fulfils [the potential] of [this] potentially hot [thing] must be [something that is] actually hot.

This *is* true in some cases. For example, it is an actual grammarian that makes a potential grammarian an actual grammarian, and fire that is actually hot changes potentially hot water and makes it actually hot. But the statement is not true in all cases. While there may be need of something actual to bring something potential to its own actuality, this actual thing need not in every case be actually what the thing that it is bringing to actuality is potentially. It is not an actual ship but an actual shipwright that makes timbers which are potentially a ship into an actual ship. And no more is it the case that the creative principles for a ship that are within the shipwright are already an actual ship.[196] [And], again, the potentially sweet, bitter or red are not in every case brought to actuality through the agency of the actually sweet, bitter or red. For the water by which plants are watered potentially has each quality into which it changes when it becomes one with the plants, and it is not, one supposes, something actually sweet or bitter or red that changes water so that it actually takes on these qualities[197] but the creative principles in nature, which are actual. And [similarly] the potentially hard becomes actually hard through the agency of the cold, although the cold is not actually hard but [only] actually cold.

And so here too he has made this kind of addition to his axiom – I mean that an actual [thing] that brings a potential [thing] to actuality must be actually of the same kind as it is potentially – so as to give himself, as I have said,[198] the opportunity of stringing together [a whole series of] creators of creators.

7. But perhaps not even a thing with a potential or a fitness for something has need in every case of another thing to bring its potentiality[199] to actuality. It is possible for someone to be himself the cause of his change from first potentiality to actuality. For according to Plato there are, as he says[200] in the *Alcibiades*, in fact two roads to knowledge: learning and discovery. So it is clear that the first person to discover an art, whether the whole of it or its parts, was himself the cause of his transition from first potentiality to actuality;[201] for instance, the first person to discover log-sawing was a potential sawyer in the first [sense of] potential before discovering it and became an actual sawyer not through the agency of someone else who was an actual sawyer but by bringing himself from potentiality to actuality. And what about a new-born puppy? It is potentially capable of sight and of barking but is not brought to actual seeing or barking through the agency of another one that actually [does these things] but comes to maturity by itself through the agency of its own nature. Unless in the first case one holds universal mind responsible for bringing potential mind to actuality, and in the second universal nature. But, even if we accept this, one would not say that universal mind is an actual sawyer or builder or anything like that, or that universal nature is actually able to see or bark, but simply that they are actual – not to prolong the argument by asking why, in the case of universal mind, it brings the potentiality of some things but not of others to actuality, and, in the case of universal nature, whether it exists at all and what it exists in. For there is not a single, universal world body over and above particular [bodies], and so there is no universal nature either (for, as Aristotle holds, the universal is either nothing or of secondary origin, existing [only] in our thought[202]), unless one were to call the concord and mutual harmony of the parts of the world universal nature.

The End of the Refutation of the Third Argument

The Fourth Argument of Proclus the Successor

The fourth [argument]: everything that comes to be from an unmoving[203] cause is unmoving in regard to its existence.[204]

56,1 For if that which produces[205] is unmoving, it is unchanging, and if it is unchanging, it produces just by being, without making a transition from producing to not producing or from not producing to producing; for if it did make such a transition, it would undergo a change – the very transition from the one to the other – and if it underwent a change, it would not be unmoving. If, then, something is unmoving, it will either never produce or always produce, so as not to be in movement by producing [only] sometimes. And so, if a thing is the unmoving cause of something, being neither never a cause nor [only] sometimes [a cause], it must always be a cause, and if this is so, it is an everlasting[206] cause.[207]

If, then, the cause of the universe is unmoving so as not to be, by moving, at first imperfect (*atelês*) and subsequently perfect (for all movement is incomplete (*atelês*) activity[208]), and so as not, by moving, to have need of time while bringing time into existence, it is necessary that the universe be everlasting, coming to be through the agency of an unmoving cause.

And so if someone,[209] with the intention of paying reverence to him who is the cause of the universe, should say that he alone is everlasting while the world is not everlasting, he is, by denying that the latter is everlasting, declaring that the former is in movement and not unmoving. And if he says that he is in movement and not unmoving, he is saying that he is not always perfect but, because all movement is incomplete activity, sometimes imperfect, and that he has need of something that is inferior [to himself] – I mean time of course – because of the very fact that he moves. And by saying that he is sometimes imperfect and not always perfect and that he has need of something inferior [to himself] he is being irreverent in the extreme. So if someone, with the intention of paying reverence to him who is the cause of the universe, should say that he alone is everlasting, he is being irreverent in the extreme.

57,1 The Sections of the Refutation of the Fourth Argument

1. An exposition of the two syllogisms in the fourth proof. Also, that because of the minor premiss common to both, Proclus has produced a fallacious argument.

2. That they often cobble together fallacious arguments that even fly in the face of their own doctrines in an attempt to subvert what we hold to be the truth.[210]

3. That that which initially does not produce but subsequently does produce, and in general, that which mobilises an activity out of a capacity, does not move or change in any way at all.

4. That the activity of God is not movement; and what is [meant by] an incomplete activity and what [by] a complete activity.

5. That not even activity that is incomplete and takes place over a period of time is movement of that which acts[211] but [only] of that which is acted upon, since that which causes the movement does not become different[212] in any way as a result of previously not causing movement and subsequently causing movement.

6. Evidence of the same from Aristotle.

7. Selections from Aristotle on the potential and the actual. Also, that it is not through[213] alteration or movement of any kind at all that the mobilising of an activity out of a capacity occurs.

8. That it is not out of ignorance that Proclus employs fallacious arguments of this kind but deliberately and to deceive the inexperienced.

9. A puzzle and its solution: that by hypothesising that God is always in capacity the creator of the world but that the world has not been created from everlasting, we are not doing away with [the doctrine] that God is always the same and unchanging. Including [proofs] that, in the case of God, capacity and activity are the same, and that it is not necessary that the things themselves (*pragmata*) should exist simultaneously with God's knowledge of things (*onta*).

10. That God's willing a thing at one time and not willing it at another is not an alteration [in him], and, speaking generally, that God's acts of will do not vary but are one and simple, always remaining the same and unchanging by virtue of always being [a willing] of the good, and that it is among the things that participate [in it] that change is observed.

11. A statement and solution of the puzzle which states that, if there has not always been a world, God was [only] potentially a creator, and therefore imperfect, before it was generated. Including [a demonstration] that every substance achieves its perfection by itself, without relation [to anything else], through the natural powers observed within it, [and] not through the actions that proceed outwards from these powers to other things.

12. That the divine is self-sufficient and self-constituting, having need of nothing outside [itself] for its own perfection. And that God has his perfection in the everlasting being of the creative principles, not in [the act of] creating.

13. That God is the creator not only of the universal but of the most particular of things as well.

14. That by the very arguments by which they think to reduce our hypothesis about the world to absurdity, claiming that if the world was generated, [God] was [only] potentially a creator and therefore imperfect before it was generated, they are led into a greater absurdity, since on their own hypotheses they make God always potentially a creator and always imperfect.[214]

15. That Proclus is once more convicted of arguing unfairly.

16. That, both according to common conceptions and according to the philosopher Plotinus, God is a perfect creator in and of himself and adds nothing to this perfection by subsequently creating, because everything is already present to God by his foreknowledge, even things that have not yet been brought to generation.

The Refutation of the Fourth Argument

1. The philosopher has included in [this] the fourth of his proofs two syllogisms, both of them with the same minor premiss, and this and this alone (I mean the minor premiss) has, because it is false, become the cause of fallacies [in the proof]. For it is a fact that, given a single absurdity, a myriad of absurdities necessarily follows.

It will be best to begin by setting out the philosopher's syllogisms themselves and then proceed to the refutation of the false premiss, that is, the minor premiss; for once this has been refuted, the fraud perpetrated at the expense of the truth by the great Proclus is undone.

The first syllogism goes like this. That, he says, which passes from not producing to producing will undergo a change, and everything that changes moves; for every real change is either a kind of movement or not without movement. And every movement is, as has been shown by Aristotle,[215] an incomplete activity. [The process of] turning white is an example. A thing which is turning white is neither black nor as yet white; for turning white is a passage to white from black. So if movement is incomplete activity and if God is not imperfect, he does not move, and if he does not move, he certainly does not change, and if he does not change, he will not pass from not producing to producing; for this kind of [transition] is, he says, change and movement. Such, then, is the first of the syllogisms.

And changing the major premiss of the above syllogism he leads the argument into another absurdity. That which earlier does not produce, he says, but subsequently produces changes, that which changes moves, that which moves has need of time (for all movement takes place in time), therefore that which earlier does not produce but subsequently produces has need of time. So if God is the producer of time, he does not pass from not producing to producing, or else he would come to be in need of time, of which he is himself the producer.

[Finally], on the basis of these [two syllogisms], he argues that if

God is perfect, entirely unmoving and unchanging, and, as being the producer of time, without need of time, and if that which earlier does not produce and then produces moves and is for that reason imperfect and has need of time, he will, so as to possess the lack of movement and change [attributed to him[216]] in the minor premiss, either never produce or always produce. So if the first [of these alternatives] is false (I mean that he never produces), he will always be producing, and if he is always producing the world, then of necessity the world too always exists. Such is the line of argument contained in Proclus' syllogisms.

2. I cannot help but be amazed at how they brush aside arguments refuting [their own position] of which they are well aware, often even in the face of correctly-argued positions of their own, and taking up arms against the truth, mislead those who are inexperienced in the subtleties of logical argument with every trick in the book.

That every movement is an incomplete activity and that every change is either a kind of movement or not without movement is true. But we should not perhaps have to argue that that which, after not producing, or, more generally, not acting, subsequently produces, or, [more] generally, acts, is not undergoing any movement or alteration. It should perhaps be enough just to cite the words in which Aristotle has taught us this, with proofs, in many of his works, words against which I think Proclus has said nothing.

But, so that we shall not seem to be depending on the opinions of others, let us first examine the argument ourselves and then in what follows put a seal on the truth by means of Aristotle's testimony.

3. That, says [Proclus], which passes from previously not producing to producing will possess change and movement.[217]

The first thing one must point out here is this: it is clear to everyone that that which previously does not produce but then produces is advancing from second potentiality, the kind that is based on [the possession of] a capacity and is also first activity, to second activity. An example is the builder. If, after initially not building, he subsequently builds, he is mobilising the activity of building out of his capacity for building. For the change from *first* potentiality to capacity is not acting (*poiêsis*) but being acted upon (*pêsis*), and that which changes from [having] first potentiality to [the possession of] a capacity is acted upon (*paskhei*) rather than acting (*poiei*).[218] For example, when the child who is potentially a grammarian is acquiring a capacity for the art of grammar, it is being acted upon by its teacher, not itself acting on someone else; and the same applies in all cases.

And so if that which initially does not produce but subsequently produces advances from second potentiality to second activity, and Proclus says that such an advance from [the mere possession of] a capacity is movement, he is clearly asserting that the advance from [the possession of] a capacity to its activation constitutes movement and change.

But if something which mobilises an activity out of a capacity moves, then there is every necessity that either the state preceding this producing or acting should be movement or that the production or activity itself or some other state conceived of as falling between producing and not producing should be; these are the only possibilities.

But it is impossible to call the state of that which is capable of producing prior to its producing, that is the state in which it had the capacity to produce without the activity, a movement. For not producing is lack of movement rather than movement, as even Proclus holds. 'If something is unmoving', he says, 'it will either never produce or always produce',[219] clearly on the assumption that not producing is a lack of movement and not a movement. Therefore the capacity to produce without its activation is not movement.

And nor is it possible to conceive of any movement that falls between producing and not producing. The mobilisation of an activity out of a capacity is instantaneous. The end of not producing and the beginning of producing occur at the same instant. A person who stops not writing, to take an example, passes to writing at one and the same moment; and the same goes for all [such transitions]. Therefore no time elapses between not producing and producing and, [more] generally, between [the mere possession of] any capacity and the activity [that flows] from the capacity. For if there is time between not producing and producing, and generally between [the members of] every [such] contradictory pair, during this [time the agent] will neither be not producing (for he has stopped not producing), nor producing (for he has not yet started producing). Therefore during the same period of time he will be neither producing nor not producing, which is impossible because [both sides of] a contradiction will be true at the same time.

And if there is no time between not producing and producing, and, [more] generally, between [the mere possession of] a capacity and an activity that proceeds from it, then, since there is no third intermediate state, nor will there be any movement, nor, [more] generally, any change, between [the two]. For every change [takes place] in time, it has been shown that there is no intervening time, and so there is no change.

Therefore that which earlier does not produce but subsequently produces, or anything at all that mobilises an activity out of a capacity, cannot move either before producing or between producing and not producing.

The only remaining possibility then is that producing itself is a movement. If this is so, and God is always producing as they claim, then he is also always moving. 'If something is unmoving', he says, 'it will produce either never or always, so as not to be in movement by producing [only] sometimes'.[220] So if he is always producing, and if producing is a movement (for this is the only part of the division that has not been eliminated), then God is always in movement. But movement is incomplete activity and in time. According to them, then, the divine is always in movement and is therefore imperfect and has need of time. So they have been trapped by their own cleverness, for this kind of fallacy is self-revealing and collapses about its own ears.

If, then, no movement is observed either before producing or between not producing and producing, and if God's actual producing is not a movement, there is no way in which God can ever be in movement when out of his capacity for producing he mobilises the corresponding activity.

But why am I talking of God? In fact, nothing else that has a capacity and then acts in accordance with it can undergo any alteration or movement solely as a result of mobilising the activity that corresponds to that capacity. For these same arguments will apply to everything, and I shall demonstrate [as much] still more clearly below.[221]

4. It is, I believe, clear to everyone that it is not right to suppose that God's producing, or his activity in general, is movement when it brings everything into existence just by willing it and has no need of time or any interval[222] to bring things into being. For it is not the case that every activity is immediately also a movement, for activity is, 65,1
according to Aristotle, wider than movement. For he says[223] that activity is of two kinds, complete and incomplete. Incomplete activity is, he says, movement. For, according to him, movement is change from first potentiality to capacity. This is how he defines it in book three of the *Physics*:[224] 'movement is the realisation of what potentially is, *qua* such'. By 'realisation' he means the actual (*autên*) activation and fulfilment of the potential. So movement is incomplete activity. By complete activity, on the other hand, he means instantaneous production from a capacity without the capacity becoming in any way different. Instantaneous production is production that does not proceed with the passage (*kinêsei*) of time but happens in an instant, like the emanation of light from a source of illumination; for as soon as a source of illumination such as a fire or the sun is visible, everything capable of it is instantaneously illuminated. Of this kind too is the activity of seeing; we instantaneously perceive sensible objects the moment we look at them. It is for this reason that Aristotle denies that the senses are in movement during the perception of sensible objects.[225] And nor is the activity of the mind movement; it grasps the objects of thought (*ta noêta*) instantaneously and without any interval. If, then, the activity of these is instantaneous, and on

that account complete and not movement, how could he[226] have the effrontery to say that the activity of *God* is movement?

5. But Aristotle does not think it right to call even activities which 66,1 are incomplete and take place in time movements of the active parties but [thinks that they should] rather [be described as movements] of the things that are acted on.[227] A mind that is skilled in building or in teaching does not itself undergo any alteration or any movement at all by building or teaching – assuming [for the sake of the argument] that the builder or teacher is perfect so that he is not learning through practice because he is imperfect. If the teacher is of this kind, and then teaches, he undergoes no alteration or any movement himself as a result of teaching. The movement that results from the teaching has taken place on the side of the pupil. He is the one who undergoes alteration and changes from not knowing to knowing. Similarly, if a mind skilled in building and [with] perfected skills, stirs a body to build, it does not become in any way different from its [former] self; the movement is seen among the building materials, namely, the stones and timber. It is they that undergo alteration; and the same applies in every case. Even though the activity in all these cases is not instantaneous,[228] the alteration and movement take place in the sphere of that which is acted upon rather than in the sphere of that which acts. For no different state of any kind arises in the mind that is skilled in teaching or skilled in building as a result of teaching or building. And even if the body moves, it is as an instrument, and its movement is different and not the kind that the things under discussion[229] exhibit; for the movement of the body is local. And [anyway] God brings all things into existence by the very [act of] willing [them] and has no need of the assistance of an instrument.[230]

6. Perhaps it will not be unprofitable to quote some passages in 67,1 which Aristotle explicitly testifies[231] to these very points.

In the third book of the *Physics*,[232] soon after defining movement, he adds the following:

> That this is movement is clear from the following: when that which is buildable is in actualisation in respect of the characteristics in virtue of which we give it this name, it is 'being built' and this is 'building'; and the same applies to learning and healing.

And a little further on:[233]

> For it is possible for each thing, as for example that which is buildable, to be active at one time and not at another. And the actualisation of that which is buildable, *qua* buildable, is [the process of] building. For it is either [the process of] building or an actual building. But when there is a building, the buildable

> no longer exists; the buildable is what is being built. So it must be [the process of] building which is the activity. And [the process of] building is a kind of movement. And the same will apply to other kinds of movement.

In these passages he is clearly saying that [the process of] building is the perfection not of that which has the capacity to build, that is, of the builder, but of that which is buildable, that is, of that which is being built, namely, stones and timber. [And] he says the same thing about learning and healing; learning is not the fulfilment of the teacher or healing of the doctor, but of the person being taught and the patient [respectively], and the same applies to every kind of movement.

And a little further on again,[234] in the course of solving a puzzle 68,1
about movement, he introduces much the same point. This is what he
says:

> The [solution to the] puzzle is clear; movement is in the moveable. It is its actualisation through the agency of that which is able to cause movement. And the actualisation of that which is able to cause movement is not something separate – for both must have an actualisation. It is 'able to cause movement' because it has the power to, and 'cause of movement' through its activity; but it is of its nature to activate the moveable.

What he is saying here is, to take an example, that even though one and the same thing, [the process of] building, is the actualisation of both the builder and of that which is being built, the movement is nevertheless not in the builder but [only] in that which is being built; for it is the fulfilment of that which is buildable that is being brought about through the agency of the builder. And the same applies to all movement.

It has, then, been shown by means of arguments drawn from Aristotle as well that production is a movement not of the producer but of that which is coming to be. So even they, constrained by the declarations of Aristotle and our demonstrations, will perhaps agree that the activity of God is not a movement. For if to create is to move, they must necessarily suppose that the divine is always in movement, since according to them it is always creating.

7. And our argument has adequately shown that anything which
at first does not produce but subsequently produces, and in general
anything which mobilises an activity out of a capacity, does not
undergo alteration or movement of any kind at all. For if it has been
shown that there is no time between not producing and producing, 69,1
and that every alteration or movement of any kind at all is in time,
there is no necessity that the creator should alter if he has always

possessed the capacity to create in [its full] perfection but has not been bringing the world into existence from everlasting.

But perhaps it will do no harm to show Aristotle himself making this very point in the second book of *On the Soul*, as we promised[235] we would at the beginning of the present chapter. Indeed, I shall begin the quotation somewhat earlier so as to present the division of the potential and the actual that we expounded in the previous chapter[236] in Aristotle's own words as well for those who are unacquainted [with them]. Here then, to quote his exact words, is what he says in the second book of *On the Soul*:[237]

> We must also make some distinctions in regard to potentiality and actuality; for up until now we have been talking of them [quite] generally. Something, then, may be a knower in the sense that we might say a man is because man is in the class of knowers and possessors of knowledge, or else in the sense in which we unhesitatingly describe a man who has knowledge of grammar as a knower. Each of these has a potential in a different way; the former because his kind and stuff are such, the latter because he can reflect[238] when he wishes, provided nothing external prevents it. (The man who is *currently* reflecting is [an] actual [knower] and has knowledge in the strict sense of [for example] this or that [letter] A). Both the first two are potentially knowers,[239] the former by undergoing alteration through learning, and frequently by changing from a contrary condition,[240] the latter in another way, [by passing from] possessing sensation, or [a knowledge of] grammar, but not exercising it, to exercising it.

That is what Aristotle has to say. And were I not intent on avoiding prolixity, I would also quote the explanatory comments of the Aristotelian commentators themselves on the subject.[241] Instead, I shall attempt a brief elucidation of the text with the aim of making the philosopher's meaning clear to those who have not attended lectures on these topics.

The circumstances of the division of potentiality and actuality are pretty clear from what has already been said by us in the previous section;[242] for since Aristotle had used 'potentiality' and 'actuality' simply and without qualification in the earlier writings we cited [there][243] (he indicates this by writing 'for up until now we have been talking of them [quite] generally'[244]), he carries out a division of them in the passage we have [just] quoted. He says that in one way every man is said to be potentially a knower, even though he may not yet have possession of the knowledge [in question], for example of grammar, but in another way, on the other hand, he says, it is the man who has already come into possession (*hexin*) of the knowledge [in question] who is said to be a potential knower. But, he says, these two are not both knowers in the same way. The former is because his kind is such. In explanation of the meaning of 'kind' he adds the words 'or stuff', by which he means that which underlies the knowledge, I mean that which is receptive of it, namely, human nature. For all men in common are said to be potential knowers through being naturally equipped to acquire knowledge; for man is 'receptive of understanding and knowledge' and is so defined, and it is this that constitutes first potentiality. On the other hand, someone, he says, who has already acquired the capacity for grammar is said to be a potential knower in the sense that he will be able to externalise his reflection[245] on his knowledge whenever he wishes, unless some outside factor prevents this. For a teacher will teach unless prevented by illness or the lack of students or their unsuitability; and similarly a builder will build, and so on. And this is second potentiality, which is also called first actuality. So when this man with a capacity [for grammar] is currently teaching and is presenting the theorems [of grammar] – this is what he means when he says 'this or that [letter] A', as clearly as though he had written 'this or that theorem [of grammar]' – he becomes an actual knower under what we refer to as second actuality.[246]

Having carried out a division [of the senses] of potentiality and actuality in this way, Aristotle goes on to argue in what follows that the man[247] who possesses the theorems [of grammar] in capacity does not, when he mobilises an activity out of this capacity, do so through [undergoing any] alteration. 'Both the first two', he says,[248] 'are potentially knowers, the former by undergoing alteration through learning and frequently by changing from a contrary condition.' A man who is said to be a potential knower in the first sense, he says, becomes an actual knower (i.e. gains possession of knowledge) by undergoing alteration through learning. For learning is an alteration; for learning stands in the same relation to someone who is undergoing change in regard to[249] knowledge as whitening does to someone who[250] is becoming white. Such a person, he says, often even changes from one condition (*hexis*) to its contrary; for through learning a learner changes either from ignorance to knowledge or from false opinion to true – and the false is contrary to the true. [Only] the latter is alteration in the strict sense. For alteration is, as Aristotle himself teaches in the *Physics*,[251] change from one form to another involving the perishing of the form out of which there is change and the coming to be of that into which there is change, as [in] becoming white or becoming hot. [For] someone who becomes white changes from black to white, which is [a change] from one form to another, the black perishing and the white coming to be; and someone who becomes hot similarly undergoes a change from cold to hot, the cold perishing and the hot coming to be.[252] On the other hand, the change from ignorance to knowledge, and that from a privation to a form generally, more closely resembles generation than alteration.[253]

That then is how someone who is said to be potentially [this or that] in the first sense advances from potentiality to [the possession of] a capacity.

But what about the second man – I mean the one who possesses a capacity and then mobilises the [corresponding] activity? He, he says, makes the transition from [the possession of] a capacity to the activity in another way. For someone who has the capacity for [using] a sense, sight for example, but has not been seeing [anything] because of darkness or because his eyes are closed or for some other reason, does not mobilise the activity of seeing because the sense has altered, but only because the hindering factor has been removed. He has not, by at first being inactive and then later active in regard to sight, changed
73,1 from one form to another or from absolute privation to the form. He had the capacity for [using] the sense even though he was not being active as far as it was concerned. So there is, he says, another way than these in which an activity emerges from a capacity.

And again – not to protract the argument by quoting the intervening material – the same [Aristotle] says a little further on:[254]

> The possessor of knowledge will reflect,[255] and this either does not amount to undergoing alteration – for it is a development into itself, or into actuality – or [only] to a different kind of alteration. For this reason it is not right to say that the possessor of wisdom alters when it uses its wisdom, any more than that that which builds does when it builds.

Observe that he states quite explicitly that something that has wisdom, that is, something that has knowledge by capacity, when it uses its wisdom, that is, when it externalises its knowledge, does not alter, and that it is not right to call this sort of thing an alteration. No change from one form to another has occurred, as happens in every case of alteration or movement; there has only been a revealing of a capacity. Aristotle was certainly making this very point when he added the phrase 'for it is a development into itself, or into actuality', which is an unambiguous indication that the mobilisation of an activity out of a capacity is not alteration. For if alteration is, as has been stated,[256] change from one form to another, the first perishing and the second coming into existence, and if that which is advancing towards its own actualisation is developing not into something else but into itself, while remaining the same thing and only revealing a
74,1 capacity by means of the actualisation, then the mobilisation of an activity is not an alteration or a movement or change of any kind at all. And should it be important to someone, he says, to apply the word alteration to this sort of thing too, this would be another kind of

alteration going under the same name and not the kind that we refer to when we talk about qualitative change. It would be like saying that night-time colour alters when, as darkness retreats, it becomes visible after previously being invisible.

But we are not discussing terminology but whether some kind of change or alteration of form really comes about in a subject when it mobilises an activity out of a capacity; and we have shown that this does not happen. And if this kind of thing is not an alteration, nor is it movement or change of any kind at all. There are three species of movement: local; in growth and decay; and in alteration. It goes without saying that anything engaging in the activity of production after previously not producing is not exhibiting either of the first two kinds of movement. It is neither growing nor decaying nor changing from place to place. Therefore that which mobilises an activity out of a capacity does not move in any manner at all.

Aristotle says the same things again word for word both in the third book of this same work and in other places, but what has been cited already will be enough on its own for those who are not wholly contentious.

8. Now that these facts have been demonstrated, how could one fail to be amazed, as I said at the outset,[257] at the trickiness[258] of the great Proclus in argument? Although the facts speak so clearly for themselves, and although pretty well everyone accepts Aristotle's 75,1
arguments both in regard to the different senses of potential and actual and [when he says] that that which mobilises an activity out of a capacity undergoes no alteration or movement, the philosopher takes no account of this – although no one would be so bold as to impute ignorance of such theories to him. [Instead] he has made it his one goal to arm himself by all available means against the truth of our Scriptures and, arguing against us as though we are novices in these matters, in the chapter before this has stitched together a fallacious argument based on the ambiguity of potentiality and activity and in the present one has dared to say that that which at first does not produce but subsequently does produce (that is to say, that which initiates the activity of producing out of the capacity to produce) undergoes movement and change (that is to say, alteration), so that he can then draw the conclusion that it is necessary either that the divine should be in movement (that is to say, should alter) or, if that is impossible – for 'with God', as the Holy Scriptures somewhere[259] say, 'there is no change of position or shadow cast by turning' – that the world should have coexisted from everlasting with the being of God.

So if it is not through movement that the mobilisation of an activity out of a capacity takes place, it is possible for there to be a world at one time and not at another even though God remains entirely unmoving and unchanging.

9. But, he claims, the very fact of saying that he at first did not produce but subsequently produced, that is, that he at first had the capacity alone but subsequently [engaged in] the activity as well as
76,1 [having] the capacity, does away with [the doctrine] that God is completely unchanging. For he will not always be the same and unchanging if he produces at one time but not at another.

Well, in the first place, my fine fellow, our own argument has demonstrated perfectly well, and the testimony of Aristotle confirmed, that the advance from a capacity to the second [kind of] activity constitutes neither an alteration nor a movement even in generated creatures, much less in God. For alteration makes its subject different, but someone who possesses a perfected capacity and then acts [in accordance with it] has not become different in any respect from his [former] self.

But perhaps what gives rise to this notion of yours, or, better, provides [you with] the material for producing fallacious arguments, is the fact that among us [human beings] those who are in possession of skills[260] also invariably have need of physical movement of the body when they want to engage in the activities that correspond to their capacities because they cannot achieve the appropriate outcomes by pure thought alone and as a result move into[261] a different state of some kind, not in the mental sphere but in the physical, I mean in the body. It is because of this that you hold up in front of the young like some kind of bugaboo [the idea] that a different state of some kind must inevitably come about in God too if he subsequently produces after at first not producing.

But if God is always the perfect creator of all things by virtue of always unchangingly possessing the principles for created things and producing all things just by willing [them] without the need of any [physical] organ to bring things into being, he will not become in the least different from his [former] self whether he produces or does not produce. For he everlastingly possesses the concepts and principles of things, through which indeed he is a creator, in exactly the same way, and does not become different in any respect whether he produces or
77,1 does not produce. For, speaking generally, it is not even proper to say that capacity and activity are different [things] in the case of God; the two are one and the same thing and difference arises in the sphere of that which shares [in them].

But I shall go back a little and try to make what I am saying clear by means of an illustration. Every skill has two aspects; one is to be observed in the state of mind which we call reflection or thought,[262] the other goes forth into the outside world, as when we communicate our thoughts to others. When we possess skills only through thinking and reflecting, we are said to possess them as a capacity, but when we project the mind's reflection into the outside world, whether by means of words,[263] as is the case with the verbal skills (we call this teaching), or by manual activity, as is the case with the practical ones, we are then said to be making active use of those capacities.[264] In the case of the verbal skills, if one were to imagine our minds on their own, stripped (*gumnas*) of [our] bodies, one would see that they no longer needed any [physical] organ to reveal their private thoughts to one another. They would make direct contact with each other's naked (*gumnois*)[265] thoughts, so that there would no longer be any difference between reflecting in the area of a capacity and actively engaging with someone else on the basis of that [same] thought, and the resulting altered state would occur in someone else,[266] [namely,] the person who is aware of the other person's thoughts.[267] And if the mind were able in the case of reflections with a practical bearing to bring the fulfilments of those thoughts into being by pure thought alone without any [physical] organ whenever it wished, then surely in that case too thinking about things and having the capacity for them would in no way differ from activity and action based on them. Since God, then, has no need of any [physical] organ for bringing things into being, but brings all things into existence by pure thought alone whenever he wills it, and since God possesses the concepts of things neither more nor less in consequence of their being or not being, it would follow that in God's case capacity and activity in no way differ.

But just because he brings all things into existence by thought alone and always possesses the concepts and principles of all things in exactly the same way, it is not therefore at once necessary that things should have coexisted with the thoughts of God from everlasting. We have also shown this in chapter two.[268] For God does not bring his creations into existence willy-nilly by a necessity of nature in the way that the sun illuminates and fire heats as soon as they are present by an involuntary necessity of nature. The cause of all things is above all necessity, for which reason it is not at all necessary that whatever is thought by God should [automatically] exist simultaneously with the thought. For it is agreed that God knows even future things that have not yet come to pass, as for example the number and nature of the souls that will participate again in this bodily life, and what kind of life each will choose, and what things will follow upon the choice of each by [the dispensation] of unerring providence. And even future time is already present through foreknowledge to the creator of time himself. Even the future revolutions of the heavenly bodies and what their relations to one another will be at any given time – for the relation of the stars to one another is different at different times – are clearly comprehended as an undivided whole by the foreknowledge of God. God will not be ignorant of these things, of which he is himself, without any intermediary, the creator. So if future things are grasped by the thought of God before they come to be, and if future things do not yet exist, for otherwise they would not be future things, there is no necessity that a thing should exist the moment God thinks of it.

And so, as has been shown, the creative principles for things [which are] within God always possess actuality and perfection but God brings each thing into existence and gives it being when he so wishes, bringing all things into existence just by willing them; and he so wishes at the time when coming into existence is good for the things that come into existence; and what is in accordance with nature is in every way good, and, as was shown in chapter one,[269] it is in accordance with nature for generated things that they not be co-everlasting with him who brought them into existence – unless they will claim that even to wish [something] at one time and not at another is alteration. And if this *is* alteration, they will be inadvertently committing themselves to the position that God is always in process of alteration.[270]

10. For, [we ask], does God want[271] each individual such as Socrates and Plato to be everlasting, or not, preferring that each of them should exist at one time and not exist at another? Clearly he does not want individual things to be everlasting, for they certainly would be everlasting if he so wished. Therefore at one time he wants an individual thing to exist and at another he does not want it to exist. And if this is so, he is always undergoing alteration. For if wishing the same [state of affairs] to exist at one time and not wishing it to at another is to alter, and if at one time God did not wish the soul of, let us say, Socrates to be together with the body of Socrates,[272] since
80,1 Socrates was not everlasting, but at another time, when he bound it to the body of Socrates, did so wish, and then once again did not wish it, when he released it from the body, then clearly, by wishing the same thing at one time and not wishing it at another, the creator has, according to them, undergone alteration. And the same argument [will apply] to all individual things. So if there are always some individual things coming to be and others perishing, he clearly always wishes some things to exist and others not to exist. But this, according to them, is change and alteration. So according to them God will always be undergoing alteration. For they will certainly not claim that he neither wishes them to come to be nor does not wish them to come to be. For [then] both sides of a contradiction will be true at the same time, which is impossible. And besides, if he does not wish them to come to be, how do they come to be? And if he does not wish them not to come to be, he clearly wishes them to come to be, for negation with reversal is equivalent to simple affirmation.[273] And further, if he did not wish them to come to be, how is it that according to Plato's *Timaeus*[274] he ordered the heavenly [gods] to turn to the creation of mortal creatures, and not only ordered them [to turn to their creation] but gave them the power to produce? And, moreover, if it is in the nature of the creative cause to exercise providence, and providence

either binds souls to bodies or releases them from bodies as is best [for them], it is clear that he sometimes wishes them to be with bodies and sometimes to be separated from bodies. And this, so these learned men believe, is alteration. [And] therefore, according to them, the divine is always undergoing alteration.

But if this is absurd, then to wish something to be at one time and not to so wish at another is not [to undergo] alteration. For [the divine] always wishes the good, or rather it *is* goodness. And the nature of generated things partakes of the emanation of the good from that source (*ekeithen*) as far as is possible for generated things.

If, then, to will something that comes about to the advantage of its beneficiaries, does not work any alteration in the divine, and if God always possesses creative power in the same measure and brings all things into being just by willing them, no alteration will be apprehended in God as a result of his producing at one time and not producing at another, for he always wills good things. So whether he produces a thing or does not produce it, it is invariably for the good; he makes each thing that comes to be as good as the nature of generated things permits. The will of God is therefore one and simple and is always the same and unchanging, for he always wills the good. It is in the sphere of things that partake of it [sc. God's will] that change and alteration are observed.

To take a similar case, the power of the sun, whether one prefers to think of its power of illuminating or of heating, is one and simple, but the things that partake of either [power] do not partake [of it] in the same manner and nor does any one [of them] always partake of the same [power] in the same manner; a bat and a man do so differently, and the same man does so differently when his eyes are unhealthy and when they are healthy, or when he has been continuously in the light and when he suddenly moves from a dark place to one that is lit up.[275]

Therefore change and alteration in the things that share in [an activity][276] will not compel us to suppose that any alteration takes place on the side of the agent (*to poioun*). And if God does not alter as a result of not producing and then producing, nor will he have need of time to produce, but will bring all things into existence at the instant he wills them.

11. Now that this last point has been demonstrated, the following [argument], which they continually trot out to the young with deceptive intent, is refuted by these [same arguments]. If, [their argument goes], the world does not always exist, before the world came to be God was [only] potentially a creator, and everything in potentiality is imperfect, so that God too was imperfect before he created[277] and became perfect [only] when he had created, for everything in actuality is, they say, more perfect than [anything] in potentiality. [And] indeed the learned Proclus himself is hinting at this very [position] in the present argument when he says 'so as not to be, by moving, at first imperfect and subsequently perfect',[278] and again, 'if he says that it is in movement rather than unmoving, he says that it is not always perfect but is at some time imperfect'.[279] And moreover in the work entitled *An Examination of Aristotle's Criticisms of Plato's Timaeus* Proclus writes, [to quote him] word for word, as follows:[280]

> If there is always a producer, there is also always something coming to be which comes to be through the agency of him who is always producing. For either not even God always produces, or he does always produce but the universe is not always coming to be, or he always produces and the universe is always coming to be. But if God does not always produce, clearly he will [at first] be potentially producing and then actually producing and [therefore] an imperfect creator and [one who] needs time [in which to produce]. And if he always produces, but the universe comes to be [only] at some [particular] time, it is impossible;[281] for whenever that which produces is actually [producing], that which comes to be is also actually coming to be. Therefore of the two the one is always coming to be, the other producing.

And how on earth, my learned friend, can something which, as has been shown, undergoes no alteration whether it is creating or not be imperfect? In what respect does a teacher become more perfect when teaching than he was when not teaching or a builder when building than when not building? Look at the matter this way. The sun is a kind of bright and luminous body. If one [hypothetically] assumed that none of the things that share in the luminous power of the sun existed (as though the heavenly bodies and the air were not translucent and could not benefit from the energy of the light), would the sun be imperfect? I do not believe that anyone would say so. The powers that are constitutive of substances do not have their being in their relation to external things. [If we did adopt such a position], we would be unwittingly accepting Anaxagoras' argument that nothing has a definite nature but all things have their existence in their relation to one other.[282] And this kind of position has been vigorously refuted by both Plato and Aristotle.[283] If, then, each substance has its being not in its relation to something else, but has the powers which constitute it within itself independent [of anything else], it clearly follows that the sun too would be complete in its own being even if none of the things that partake of its illuminative power existed. Therefore, even should they partake [of it], it will not be the least bit greater in regard to its constitutive powers. And the same applies in the case of the heating power of fire: even if there is nothing that is being heated, fire is nonetheless complete under the definition of its essence, and, conversely, when something *is* being heated, it will not be any more

complete, since heat is not present to fire or the power of illuminating 84,1
to the sun relationally; for powers which are constitutive of substances, [those] without which [their] substratum could not survive, inhere in things independently and without relation [to anything else]. Even at the bottom of the sea a shell has the principle of colour in its entirety, even though it does not come under observation.[284] And if these things are so in the case of corporeal things, how can someone who does not even concede as much to the creative cause of existing things and does not want its essential powers to be independent of external relations be other than truly blasphemous and egregiously impious? For just as fire is hot and the sun bright by nature and of themselves but they heat or illuminate in their relation to external things, so assuredly is God always a creator by virtue of his perfect possession of creative principles (as is a builder even though he is not building), whereas he creates only when he also directs his own activity outwards, no longer being constituted by it. For everything in existence is characterised not by the activities that proceed from it but by its essential powers. Man is capable of speech by virtue of having the capacity and it is by virtue of this that he has his being, not by virtue of speaking; if he does not speak, he is nonetheless capable of speech, and snow is cold even if it is not cooling [anything] because there is nothing present to be cooled. And the same applies for each and every activity.

But if they persist in claiming that God can only be perfect if he has created the world from everlasting, let them answer one small question for us.

12. Is it or is it not the case that the divine is self-sufficient and 85,1
constitutive of its own power and substance? If it is not self-sufficient – let the blasphemy be on the heads of others! If it is self-sufficient and suffices for itself so that it is in want of nothing and perfect and stands in need of nothing outside itself – for everything that has been brought into being by the divine substance through an act of creation is completely external to it and removed from any essential relation to it – then God will be perfect, being perfected through his own agency even should none of his creations exist.

Another argument. Either the perfection and absence of need of the divine substance are present in it from within from its own resources and have need of absolutely no external impulse, or they are supplied from some other source. If they [are supplied from] some other source, the supplier is more blessed than the recipient and greatly superior and more free of need. For how could it have given what it did not have? So what is this [other source]? What else, according to them, than created things, without which, they do not fear to claim, God would not be perfect? For if God cannot be perfect unless created things also exist, then his products will be perfective of the producer himself. Such, then, is the situation if perfection has come to God not from his own substance but from outside. But if [it comes] from nowhere outside – for this is taken for granted even by those with the most inadequate[285] ideas about God – and his absence of need and perfection have come to God from his very substance, but 86,1 he could not be perfect unless things brought into being by him are always coexistent with him, then all things are part of the divine substance, even the very least of things, matter. For things that are constitutive of substances are parts of substances. For if it was agreed earlier that perfection is present in God from his very substance, and if, on the hypotheses of the Hellenes, God could not be perfect unless the world existed together with him from everlasting, and if things which are constitutive and perfective of substances are parts of [those] substances, as the rational faculty is of man and the hand [and] the foot of our body, then the world will be part of the substance of God. They can work out the consequences of this themselves. For, if it is so, God will be body and will be affected by all that goes with bodies.[286]

Another argument. If perfection could not be present in God unless things have existed along with him from everlasting, God will get back from the things he creates more than he gives them. He will give them being but get back from them perfect being, and it is much better to be perfect that simply to be. And so God will create things not as a result of his own goodness, so that there will be things to share in it, but so that he can lead a life without pain or need with the cause of his perfection always alongside him, just as we, for instance, for our own protection, produce dwellings and clothing and the like, not so as to benefit these things, but providing for our own comfort.

These learned men then have, by means [of the very arguments] by which they intended to ascribe perfection to him, inadvertently 87,1 represented God as imperfect and not himself sufficing for his own perfection but dependent on things that are brought into being through his own agency for his existence.

But perhaps someone will say to this: 'We are not saying that it is the [created] things that contribute to God's perfection, but rather the activity creative of all things which is present in God as part of his essence, of which the existence of things is an inevitable consequence.' But no less absurd consequences follow from this too. If his activity is imperfect if nothing has come into existence, and if God is not perfect if his activity is imperfect, then his perfection has come to him from created things via his activity. But, men of wisdom, activities in relation to external things are not perfective of those who act. For this reason Aristotle himself says[287] that a teacher, and in general anyone engaged in any activity at all, does not undergo alteration as a result of externally directed activity, since a thing that is being perfected has something done to it and is altered, but, as we have often shown, one who acts has nothing done to him and undergoes no alteration. And so even a teacher is not being perfected when he teaches. And if teaching is not perfective of the teacher's [already] perfected capacity, neither will creating be constitutive of the creator's substance. Only his creative capacity or power itself will be, lest the being of God should lie in his relation to created things. For if creating is constitutive of the divine substance, and if creating has its being in relation to created things – for he who creates creates by being the creator of created things – then the divine substance will find its completion and perfection in its relation to created things. And this being so, it would have need for its existence of things to which it has itself given being. And what could have a more sacrilegious effect on our notion of God than this?

So if thinking or saying this sort of thing about God is the extremity of impiety, and if the statement 'if the world does not always exist, God will be imperfect before the world comes to be' has led to this, then God is perfect even when none of the things brought into being by him [yet] exists.

13. It is also of value to look at the matter this way.

The doctrine that the origin of things is single compels belief that God is the cause of all things from the first to the last. 'All things', says Plato, 'exist in relation to the king of all, and all things exist on his account, and he[288] is the cause of all beautiful things.'[289] And he further says that God embraces the beginning, the end and the middle of things.[290] 'The rule of many is not a good thing; let there be one ruler', as both Homer and Aristotle hold.[291] For even if some of the things that come to be come to be through the agency of certain intermediate causes, it is nevertheless quite clear that what is most of all their cause is the primary cause of their existence, which also furnishes the intermediate causes with the power to be causes.

That the primary cause *is* more a cause than the immediate cause is clear from the facts themselves, if it has, as I said,[292] even furnished the secondary causes with the power to be causes. But there is nothing like [being able] to cite Aristotle to exactly the same effect. For in the eighth book of the *Physics*, after saying[293] that something which is moved is moved by a mover and that the mover either moves [it] directly or through [one or] more intermediaries, as when one moves a door with a stick – where the first mover is the soul and the last the stick, for the soul moves the hand, the hand the stick, and the stick the door – after saying this, he concludes as follows:[294]

> We say that both these [earlier movers] and the last impart movement, for they move the last but not on account of that the first, and without the first [mover] the last will not impart movement, but the first will without [the last]: the stick, for example, will not impart movement if the man does not impart movement [to it].[295]

And again a little further on[296] he says:

> For it is clear that everything that is moved is moved by a mover that is earlier [in the series], and more so by the first of these movers.

And again a little further on:[297]

> There will be no first mover if each of the two [parts] is to move itself; for that which is earlier [in the series of movers] will be more the cause of the movement than that which follows it and will more [truly] impart movement; for there were two ways to impart movement, (1) by being oneself moved by something else, and (2) by oneself; and that which is further from the thing that is moved is closer to the source [of movement] than that which is intermediate.

In these passages he clearly states that even if there are a number of intermediate movers it is, as is the case in the example he has provided, the first cause of the movement that is more a cause[298] of the movement. For the soul is more the cause of the door's moving than is the stick, and for this reason he rightly says that the more remote mover is closer to the thing that is moved, as the soul is closer to the door than is the stick, not in place but through its power and by being more the cause of the movement.[299] Since, then, the soul is the first cause of the movement, he says that the soul is closer to the door than is the stick.

And what Aristotle says about movement is also clearly true of every cause. Even should a teacher teach using an actor as an intermediary, the teacher is more the cause of the [the pupil] being taught than is the actor who plays the part of a teacher.

And the same applies to generation. The first cause of generation is more a cause than the more immediate causes. Even though the father is a cause of generation, God, who sows the creative power in nature, is clearly more the cause. (Even if they do not want the creator to be the first cause, since their 'first god' is the cause of the creator's being a creator, he in turn would be the chief cause of the existence of all [other] things as well.)

So, in view of these proofs, there is every necessity that the first god should also be the cause of the generation of particular things, even if they assume ten thousand intermediate causes.

And moreover in chapter two we cited passages from Proclus' own work *On Ten Puzzles About Providence* in which Proclus clearly states that the creative and provident power of God extends to the meanest and most insignificant of things. Here is what he says:

> In summary, then, we claim that this unity produces all things and sustains all things and that it has an existence more real than all being and clearer than all knowledge.[300]

And again:

> But it [sc. the unity of providence], remaining stationary and undivided in [its] oneness, also knows all things in this same mode, and not only man and sun and everything whatsoever of that kind, but also each particular thing. For nothing escapes that unity, whether it is a question of its being or of its being known.[301]

And again:

> Thus the unitary knowledge of providence is, in the same undivided [entity], the knowledge of all divided things, both of all that are most individual and of all that are most universal; and just as it has caused each thing to exist in unity, so does it know each in unity.[302]

From all of this, then, it is clear that the creative power of God necessarily reaches to the most individual of things.

14. Since, then, God is the cause of the existence of the particular 92,1
and most individual, such as Socrates and Plato and this horse and this plant, let them tell us whether or not God is always the actual creator of all individuals. If, on the one hand, he is always[303] the actual creator of all things, how is it that all things are not present at the same time, but some have perished, some exist [now], and some are going to [exist in the future]? And if he is the actual creator of future things, how is it that future things are not [now] present? But if he is a potential [creator], and the potential is in their view imperfect, then God is in their view always imperfect, if it is always the case that of individual things some perish and others are yet to come into existence. And if, as the learned Proclus holds, [it is the case that] if the world exists potentially, its creator too will exist potentially (as he argues in the third proof[304]), and because of this there is need of another [creator] to bring him to actuality, and if, on the one hand, the second one is actual, the first one too must be an actual creator, but if the second too is potential, there is, he says, once more need of another, third, [creator],[305] and so on *ad infinitum*, [then] it certainly follows on the same hypothesis, if[306] the proximate cause of future individuals is[307] a potential cause, that the cause of the proximate cause is potential as well, and making our way in this fashion all the way to the first [cause], we shall be compelled to say that it too is potential; for if the first cause is the actual cause of the existence

of the second, the second too will be an actual cause, and if this is so, the third in its turn will be an actual cause, and so on until the last
93,1 cause, just as, if the last cause, the father for example, is potential, the first too will necessarily be potential. Since, then, some individuals are *going* to come to be (I mean men and plants and all the rest), the proximate cause of their generation is clearly a potential cause; and so the cause of their cause too will be potential, and likewise all the way to the first [cause]. And everything potential is, according to them, imperfect. So, according to them, the first cause will always be potential and imperfect. For even if God is the actual creator of things that have already come into existence, he will certainly be the potential [creator] of future things. So either one must admit that potentiality based on [the possession of] a capacity is perfect or God will certainly always be an imperfect creator of future things; for he will always be a potential creator of future things.

And this must be the case not only in the creation of particulars but of universals too, of which God is the creator without any intermediary; for, as Plato says, the creator brought the heaven and all the species into being without any intermediary. At any rate, in the *Timaeus,*[308] after the generation of man, the creator spoke to the intelligible and perceptible gods of the heaven, the sun and the moon and the rest, to this effect. Here is the passage:

> When all of the gods, both those who revolve unseen and those who are visible when they so will, had come to be, he who fathered this universe addressed them as follows: 'Gods, off-
> 94,1 spring of gods, works of which I am the creator and father ....'[309]

And a little further on:[310]

> 'Listen now to my instructions. Three kinds of mortal beings still remain ungenerated. If they do not come to be, the heaven will be imperfect, for it will not have within it every kind of living creature, and it must if it is going to be fully perfect. But if they were to come to be and receive life through my agency, they would be equal to the gods. Therefore in order that they may be mortal and that this universe may truly be a universe, turn, in accordance with your [respective] natures, to the creation of living creatures, imitating the power I used in your generation. I shall make a beginning by sowing in them the part of them which is called divine (in so far as anything in them is appropriately named for the immortal gods), and the guiding principle, [and then] I shall hand them over.'

If, then, everything in the heaven, and in general everything that is in their opinion everlasting, is brought into existence by the creator without any intermediary, is God always bringing them into existence, or has he stopped bringing them into existence, having created them once and for all, thereafter [only] preserving and maintaining them?

Now, if he has stopped bringing them into existence, he definitely also began; for everything that has ceased must certainly also have begun. But if he is always creating – for this is what they believe, as Proclus himself has stated a number of times in what has preceded; in the third proof,[311] for example, he says: 'Now, if the creator is always a creator in actuality, that which is being created will likewise always be [in the process of] being created in actuality'; and again in the same [proof]:[312] 'it follows that its effects are always in a state of actualisation too and that the world is always [in the process] of being 95,1 created'; and he says the same in the fourth and fifth [proofs], and in the recently cited[313] work which he wrote against Aristotle's interpretation of the *Timaeus* Proclus says, [to quote his] exact words, 'the world is always [in the process of] being created', and 'just as the creator has always created and is [always] creating, so has the world always been created[314] and [so] is it always being created, and [it is always] coming to be [although] it has [already] come to be, and although it has [already] come to be, it is [nevertheless] always coming to be', and, a little further on, 'and so the world is always being created, and just as [the creator] has always created and is [always] creating, so has the world always been created and [so] is it always being created, and [it is always] coming to be [although] it has [already] come to be, and although it has [already] come to be, it is [nevertheless] always coming to be' – if, then, this is true, I am amazed that he has not seen that he will be hoist with his own petard.[315] For if God has both created the world already and is creating it now, and for this reason the world for its part has already come to be and is coming to be now and will come to be in the time to come because it does not have its entire existence present to it all at once as eternal things do but parcelled out along with time, it is, one assumes, necessary that the [world] that is coming to be now should, in so far as it is [in the process of] coming to be, have existed potentially before being present in the here and now. For if it had actual existence earlier as well, it did not have actual existence in respect of its present existence, but in respect of its having come to be and existed then; for if it had actual existence earlier in respect of its coming to be and existing now before being present in the here and now, the world would not always be in the process of coming to be and its being is not divided up along with the division of time, but it has its existence all at once as eternal things do.

But not even Proclus believes that this is the nature of the existence of the world. 96,1 At any rate, in the work that has [just] been cited, he says:

> And if Timaeus calls the world 'a god that is going to exist at some time'[316] – [which] could perhaps cause some to worry that he may be attributing generation within a part of time to it, for 'some time' is a part of time – we must say that everything that is in time, whether infinite or finite [time], will always exist at some [particular] time; for as much of it as there is [at a given time] is in a particular [part of] time;[317] for time does not exist all at once but bit by bit.[318] So if something is in time, even if it lasts for an infinite time, it [always] exists at some [particular] time – but it comes to be *ad infinitum*, not at some [particular] time,[319] always changing from one [particular] time to another [particular] time; and it has existed at a [particular] time [in the past], and it exists at a [particular] time [now], and it will exist at a [particular] time [in the future]; and this '[particular] time' is always a different one, but nevertheless it is *in* the [particular] time it is in. So that which has its existence in a part of time comes to be at some [particular] time and exists at some [particular] time and will exist at some [particular] time; but that which [has its existence] in the whole of time exists at some [particular] time but is always coming to be, mimicking by always coming to be that which always is.

If, then, as Proclus himself believes, the world, on account of its being in time, always exists at some [particular] time (because[320] as much of the world as is in existence exists in some part of time, since the whole of time does not exist at once but [only] bit by bit), and [if] it [sc. the world] is always passing from one [particular] time to another [particular] time, and just as it *was* at a [particular] time and *is* now at a [particular] time, so *will* it be at a [particular] time, and
97,1 what will exist at a [particular] time, in so far as it is *going* to exist at a [particular] time, does not now as yet actually exist, since neither does the part of time that is going to exist at some time [in the future] actually exist now (for [otherwise] the future would already be actually present), [then] it is, one assumes, necessary that the existence of the world that is to occur at some time [in the future] should be a potential [existence] now. And likewise therefore he who is the cause of the future existence of the world is its potential cause now; for just as the fact that the world already actually exists at a point in time does not prevent it from having potential existence, since, because its existence is parceled out along with time, it is going to exist at some time [in the future], in the same way the fact that he has already actually been its cause, in that it has already existed, does not prevent him who is the cause of the existence of the world from being so potentially in that he will be the cause of its being again; for that which produces must have the same status as that which comes to be, and, conversely, that which comes to be as that which produces, as

Proclus himself has often said[321] in what has gone before; for if, because the world will exist at some time [in the future], God is now its actual cause, and when that which causes and produces is actual, that which is caused and comes to be will also be actual, as we have recently heard Proclus say[322] in the passages we have just quoted, then even the existence of the world that is to occur at some time [in the future] will [already] be actual now; and so the future time in which the future existence of the world will at some point occur will also be actual now; [and] therefore the future is already actually present, which is ridiculous and impossible. Therefore it is not possible for God to be the actual creator of the world now on the ground that it is going to exist at some time [in the future]. Therefore if, 98,1
because the world is going to exist at some time [in the future], it does not yet exist actually but only potentially, then he who is the cause of its future existence is, in relation to its future coming to be, now its potential cause, since God is also in the here and now the potential cause of future time. And there will always be [future] time and the world will always be going to exist at some [future] time, so God too will always be the potential cause of both time and the world; for just as the fact that he has already built does not prevent a builder from being the potential builder of things that are to be built in the future, but in so far as he has built he is an actual builder and in so far as he is going to build a potential [builder], so too, one assumes, is God, in so far as he has brought the world and time[323] into existence, their actual cause, but in so far as the world and time are going to exist at some time [in the future], he was and is their potential cause; and the world and time are always going to exist at some time in the future, so God will always be their potential creator.

So, because of the expert nature of their reasoning and the unavoidability of its truth, the more inferior and imperfect it is to be always potential than to be so [only] on one occasion, the deeper does their hypothesis of the continuous generation of the world fall into absurdity.

15. It is, I believe, worth looking closely at Proclus' villainous way with arguments in the words we have just quoted[324] as well.

Having seen where the argument is heading, I mean, of course, to the [conclusion] that the creator is always potential, he attempts to mask this by the ingenuity of his argument, but even so he is readily exposed by the truth. For having said[325] 'everything that is in time, 99,1
whether infinite or finite [time], always exists at some [particular] time, because as much of it as there is [at a given time] is in a particular [part of] time,[326] for time does not exist all at once but bit by bit' and again[327] 'if something is in time, even if it lasts for an infinite time, it exists at some [particular] time' and[328] 'it is always moving from one [particular] time to another, and it existed at a [particular] time, and exists at a [particular] time, and will exist at a

[particular] time' and[329] 'the "[particular] time" is always a different one, but nevertheless it is *in* the [particular] time it is in' – having made these statements, and seeing, as I said,[330] that on his hypotheses he who is the cause of something that is going to exist at some time in the future must, in so far as it is going to exist at some time in the future, be its potential cause in the here and now (certainly not its actual one, if it is indeed impossible for something that is going to exist, in so far as it is in the future, to be actual in the here and now), wanting to mask this he adds that it exists at some [particular] time, but comes to be indefinitely rather than at some [particular] time, and [says] next[331] 'that, then, which has its existence in a part of time comes to be at some [particular] time and exists at some [particular] time and will exist at some [particular] time, but that which [has its existence] in the whole of time exists at some particular time but is always coming to be', so that by saying that it comes to be[332] always and not at some [particular] time he could avoid, or so he supposed, saying that the creator of that which comes to be is a potential [creator].

And yet, if the world does indeed always exist in a continuous and uninterrupted [state of] generation and has its being in coming to be, if it comes to be not at some [particular] time but always, it will exist not at some [particular] time but always; for its being has its existence in its coming to be. But if, because the world exists in time and does not have the whole of its appointed existence present to it all at once as eternal things do (because neither does the whole of time exist at once as eternity does), the world is for that reason always in a [particular] time, and was at some [particular] time and is at some [particular] time and will be at some [particular] time, always moving from one particular time to another particular time, and its being is the same thing as its coming to be, it necessarily follows that just as it was at some [particular] time and is at some [particular] time and will be at some [particular] time, so did it come to be at some [particular] time and so is it coming to be at a [particular] time and so will it come to be at a [particular] time; for it only participates in being by coming to be. So if its coming to be is the same thing as its being or if its coming to be is the cause of its being, there is every necessity that its coming to be should have the same characteristics as its being – should it be the same thing, for that very reason, and should it be [its] cause because they too believe that a cause has the same characteristics as its effect, and, conversely, an effect as its cause. And so if the world exists at a [particular] time it necessarily also comes to be at a [particular] time.

And quite apart from its coming to be, just to say that the world will exist at some time [in the future], unless it is going to exist without a creative cause, renders it necessary,[333] on their own hypotheses, that the cause of the future [world], since it *is* future, be potential now. So if it is the case that the world is always going to exist in the future because time is also going to, it is necessary that the cause of that which is always going to exist, in so far as it is *going* to exist, be a potential [cause]. And so, even without saying that the world will be coming to be in the future, the argument is, thanks to their own irrefutable reasoning, reduced to the same absurdity, I mean to [the conclusion] that God is always a potential creator and always imperfect. [And] therefore not even by ingenuity has Proclus been able to obscure the absurdity of their hypotheses.

16. If, as pious conceptions of God have it, God himself is his own 101,1
plenitude and he has need of none of the things[334] he has brought into existence for his perfection, it is clear that he does not derive anything that contributes to his perfection from the existence of things and that he will not be any less perfect than he is[335] if they do not exist, if,[336] as Plotinus holds, even things that do not yet exist are present as existing things to God. In fact, we earlier[337] quoted a passage of his from his treatise *On the Intelligibles, or, On the Good,*[338] which is worth looking at again now. In it, speaking of God's creative power, he says:[339]

> But surely if it is necessary that each activity [of his] should not be incomplete and it is not right to think of anything belonging to God as other than whole and entire, all things must be present in anything of his. So existing for ever must be too. So the future must already be present too. There is no later in that [place]; what is already present there comes to be later elsewhere. If, therefore, the future is already present, it is necessarily present as having been provided for with a view to later [time]; that means that nothing will need for anything then, that nothing will be left out. All things, then, already existed and always existed and existed in such a way that one could say later that this exists on account of that. When it is extended and, as it were, unfolded, it can reveal this after that, but when it is together, it is all 'this'; that is, it has its cause too within itself.

If, then, as the philosopher Plotinus holds, that which is going to exist but has not yet come to be is already present in God and nothing is
later as far as he is concerned, but what is going to exist later is 102,1
already present for God and is always present and is present as having been provided for with a view to later time, and if because of this every activity of God is perfect in that it grasps what is going to exist later through foreknowledge, so that God's activity does not fall short in any respect or need for anything when it eventually brings into existence what was previously in the future – if these things are true, they must, one assumes, accept [that] the same things [are true]

even if the world has its existence from some beginning. And so, clearly, even before the world existed it was present to God as having been provided for a later time. And for this reason God's creative activity is perfect in that nothing is later as far as God is concerned, because what is going to exist later is already present to him through foreknowledge, and in that he does not need to add anything else later for the production of the world.

So since, on this showing, even they, in conformity with [our] common conceptions about God, accept all of this, they can only be entirely without shame and in contradiction of sound doctrine of their own when they say that, unless the world is everlasting, its creator will be imperfect before it comes to be, and that, when he creates it, he will need time and will not be exempt from alteration. God, then, since he is his own plenitude in every respect, will not make use of anything arising from an external relation in perfecting himself.

The End of the Refutation of the Fourth Argument

103,1

The Fifth Argument of Proclus the Successor

The fifth and next[340] [argument]: if time exists[341] together with[342] the heaven and the heaven together with time, there is no heaven if there is no time nor any time if there is no heaven.

But there was no 'when' time[343] did not exist and there will be no 'when' it will not exist. For if there was a 'when' time did not exist, time, as it seems, existed when time did not exist; for what exists at some time[344] is said to exist at some time because it does not exist at another time, existing neither always nor never, but [somewhere] between the two; and wherever there is 'some time' there is time. And if there will be a 'when' there is no time because [time] has changed from being [in existence] at one time to not being [in existence] at another time, then, since time will not exist, there will then be time when time does not exist; for 'some time' is temporal.[345]

So if there neither was a 'when' there was no time nor will be [one] when there is none – for in either case, even though there is no time, there will be a 'some time', which is [a part] of time, in existence – then there is always time; for either ['always'] or 'never' will be opposed to 'at some time'; it cannot be 'never', for time certainly exists; so there is always time.

And the heaven *does* exist together with time and time together with the heaven; for [time] is the measure of the movement of the heaven, as eternity is of the life of the living-creature-itself. (In fact, this of itself shows that time always exists, [for it must be so] so that it will not be the case either that eternity, though remaining eternity, is not a pattern for anything because there is no time, or that it does not even itself have the [characteristic of] always remaining what it is because it changes from not being a pattern to being one or to not being one from being one).

So the heaven, like time, exists for ever,[346] being born together with 104,1
time and having come to be neither before nor after time, but, as [Plato] says,[347] having existed and existing and going to exist for all of time.

The Sections of the Refutation of the Fifth Argument

1. That Proclus sophistically highlights[348] the inadequacy[349] of the words [employed] rather than the meaning of what is said.

2. That the time expression 'some time' does not in every case and in every context[350] indicate a segment of time.

3. That nor are the other time words always and everywhere indicative of time.

4. That even if we say 'there was some time when time did not exist and there will be some time when it does not exist', we are understanding [the words] 'was' and 'some time' appropriately to the nature of the subject and intending them to be indicative not of a segment of time but of a sort of existence[351] of eternity, in which there neither was nor will be time.

5. That even if eternity were the pattern for time, there [would be] no necessity that time should always exist as eternity always exists.

The Refutation of the Fifth Argument[352]

1. This proof too is excessively sophistical. He has directed his response not to the sense of the text but to its wording. If, he says, there was a 'when' time did not exist and there will be a 'when' it does not exist, there was some time when time did not exist[353] and there will be some time when it does not exist. But, he says, 'some time' is indicative of time and wherever there is[354] 'some time' there is time.
Therefore there was time when time did not exist and there will be 105,1
time when time does not exist.

Thus argues the philosopher, even though he is not unaware that words are not wholly able to image pure thought. What, then, if, to evade his over-close scrutiny of the words, rather than say 'there was some time when time did not exist' or 'there will be some time when [time] does not exist', we simply [say] that there is not always time but it has a beginning and an end to [its] existence? And if, when we say this, the philosopher, so as to drag in the words 'some time', argues that what does not always exist must, because it neither always exists nor never exists, exist at some [particular] time and not exist at another time, let him allow us to imitate his sophistical manner [of arguing] and engage in a little wordplay [of our own]; for

Homer in his wisdom rightly holds[355] that a man gets his own words back. So let him evade his own sophistry as best he can.

2. Given, then, that part of time has passed and part is in the future, will a future part of time such as tomorrow exist at some [particular] time or not? For if tomorrow does not always exist, it will exist at some [particular] time, since, to use [Proclus'] own words,[356] 'either "always" or "never" will be opposed to "at some time" '. So if both of these alternatives are wrong (for what is in the future exists neither always nor never, since it would no longer be in the future), there is every necessity that it should exist at some [particular] time. So if 'at some [particular] time' is indicative of time, and if the future part of time is also time, then there will be a time when there is going to be time. [And] therefore there [will be] a time in which future time
106,1 such as tomorrow will exist. And the same puzzle arises again in regard to that [time]; for it too will exist either at some [particular] time or always or never. And since 'always' and 'never' are wrong (for 'some [particular] time' refers to a part of time), some [particular] time will exist at some [particular] time; but some [particular] time is time; so this time too will be in time, and so on *ad infinitum*.

But in my opinion arguments, or rather fallacies, of this kind will be better suited to boys at play, who, with little interest in the truth or meaning of what is said, pull statements apart and manage to obscure the drift of the argument by babbling on about its components,[357] rather than to men who claim to have some common sense; for one should focus on the sense of what is said even when the weakness of the language fails to achieve the accurate expression of what is meant. So just as when we say here [in this present statement] that a future part of time such as tomorrow will exist at some time, we are not by [using the words] 'at some time' introducing any [particular] time, even though [the word] 'at some time' is indicative of a temporal signification, but only indicating that the future does not *always* exist, in the same way, one supposes, even if one says that time does not exist at some time or will not exist at some time, one is not in saying 'at some time' referring to another time in which time did not or will not exist but only indicating by 'at some time' that [time] neither always exists nor never exists; for in terms of existence 'at some time' will be midway between 'always' and 'never'. For if when one says that time does not exist at some time 'at some time' is indicative of time, then clearly when one says that the future segment of time exists at some time, 'at some time' also signifies time; for to exist at some time is the positive of to not exist at some time. So
107,1 whatever 'at some time' may mean in the affirmative it will also necessarily mean in the negative. So if when one says that [some part of] the future, such as tomorrow, exists at some time and not always, 'at some time' does not in itself refer to any specific time, then neither when one says that time does *not* exist at some time is 'at some time'

indicative of any specific time; for just as in the case of a particular time like tomorrow, which will exist at some time and not always, 'at some time' does not refer to time, in the same way in the case of time as a whole, which is said to exist or not exist at some time rather than always, 'at some time' of necessity does not refer to time; in either case all that is meant is that the time of which 'at some time' is used does not always exist.

So how can it be anything other than ridiculous, given that it is not possible to refer to things that have a beginning and an end to their existence other than by means of time words of this sort, that a philosopher who is eager to hold to the truth alone should not focus on the meaning of the argument but attack its wording?

3. And it is clear from the following considerations that the use of this or that time word does not in every case at once introduce the idea of time.[358]

If someone should say 'when something in movement stops[359] moving, then it also begins to be at rest', has the speaker immediately also indicated a time just because he has used the time words 'when' and 'then'? That would be impossible. The moving object stops moving and starts being at rest at one and the same instant, for neither the end of the movement nor the beginning of the rest is in time. A thing whose nature it is to be at rest or in movement must do so in time, but what is under discussion is neither movement nor rest but the 108,1
end of movement and the beginning of rest, [and] therefore 'when' and 'then' are not here indicative of [a period of] time but indicate the end of the time in which the movement has taken place and the beginning of the succeeding time in which there will be rest.

So when time expressions are used they do not in every case at once indicate time. So why should it not be the case in the present instance that if the words 'there will be some time when time will not exist' are uttered, 'some time' should refer not to [a period of] time but to the end of time, with which time will cease existing? And if time is said to have come to be 'at some time' [and] not to exist from everlasting, that [it] should in that case refer to[360] the beginning of time and the first instant or moment with which it began to exist?

And, indeed, in the same way 'was' and 'will be' do not always involve a reference to time in every context where they are used. The verb 'is' is in the first place indicative of existence, but there is also a concomitant indication of time – just as there is a concomitant indication of a segment of time in the case of verbs with other meanings – but this certainly does not mean that 'is' and 'will be' will always indicate time in every context where they are used. If someone were to say 'even before all men, or even any of them, knew God, God was', does 'was' as predicated of God here refer to time or some aspect of time, such as its beginning or end? That is impossible; the divine is not in time. So the verb 'was' is here indicative only of God's existence.

For if 'was' is always indicative of time, then, since it is indicative of 109,1 past time, and the past does not exist, one must suppose that, as far as the learned Proclus is concerned, when Plato says[361] of God 'he was good, and in the good no envy ever arises in regard to anything', he has located God in time, and in past time at that, and that therefore he is either no longer God or no longer good or [his existence] is divided up along with the division of time and not all at once and eternal. For if 'was' is predicated of something, 'is' is not predicated of it; if someone says 'Socrates was', he is clearly saying it of someone who does not exist now. For this reason one cannot just simply say 'the heaven was' because it still is. Similarly, if a man has a good disposition, it is not possible to say of him without qualification 'he was good', but only 'he is good', whereas of a man who is no longer good but has been converted to vice one may say 'he was good' because he is now clearly no longer good; for even though it is said of things that have their existence over a period of time and still exist that they both were and are – of the heaven, for instance, for the heaven both was and is – nevertheless in so far as they were, they clearly are not now. The existence that was theirs in past time now is not; if it were, the past would be present. And so, in so far as 'was' is predicated of something, it does not exist now as far as that individual existence [it had then] is concerned. So if when Plato says of God 'he was good' he is not locating the divine in time or dividing [its existence] up along with the segments of time, then 'was' does not always indicate time in every context.

110,1 And, besides, there is every necessity that they either concede that among intelligible and divine entities there are some that are supramundane and free of any relation with bodies – especially[362] God himself, the first cause and origin of intelligible entities and of simply everything – or [that] they will deny that any intelligible entity is divorced from all association and relation with body. Now, if none of the things that exist, not even the very first of them, exists with its substance and being free of any relation with body, they could not even exist without body; for if bodies were eliminated, things whose existence lay in a relation with bodies would necessarily be eliminated along with them; and in that case bodies would contribute to their being and be in a sense the cause of their existence. But if this is absurd, then there is every necessity that the intelligible substances, and immeasurably more so God, the cause of all things, should have their being above and beyond any relation to bodies and the world. But it is certainly agreed by all[363] that time is nothing other than the number of the movement of bodies and that, in a word, it has its being in the ambit of bodies; for years and months, days and nights, which are parts of time, have no being apart from bodies. And if time is the number of movement, it clearly takes third place, behind bodies; for all natural movement is the movement of bodies, and time is a measure of movement, or simply something [associated with] movement. [And] therefore the things that transcend [all] relation to bodies are, one supposes, very much further removed from [any] relation to time, which comes third, after bodies. So if this is true and we can say in regard to intelligible entities and in regard to God that God was even before this or that came into existence, and if not only the everyday speech of men but the entire usage of philosophers is full of [this way of talking] (for there are occasions when we cannot express ourselves in any other way because the expression of the subject matter demands it), then 'was' does not invariably introduce [the idea of] time in all of its usages and in every context. After all,[364] we also say 'there was a time' and 'was' is certainly not indicative of time there; otherwise there would be two times [in existence] at once, the time of which we predicated 'was' and the 'was' itself, which is absurd. And there will even be three times in existence at once if someone says 'there was a time when Socrates was', and each of the two instances of 'was' is separately understood temporally.

But I imagine that this sort of thing must seem ridiculous to everyone. One could contrive [to have] four or five times at once by using 'was' that many times in the same statement. If, therefore, 'was' in these examples is clearly indicative only of the existence of the things it is predicated of, and not at all of time, then 'was' will not invariably indicate time in all of its usages and in every context.

And, indeed, in the same way nor do the words 'will be' always in themselves signify the future segment of time in every context, and therefore, should we happen to say 'there will be a "when" there is no time', it is [only] by interpreting 'will be' temporally[365] that they [can] draw the conclusion that there will be a time when there is no time. For[366] if someone declares that 'just as there was time and there is time now, so too will there be time [in the future]', and then someone [else], using this ingenious mode of argument, interprets the 'will be' temporally, he will obviously draw the conclusion that there will be a time when there will be time.[367] For if it is true that time will be, and if 'will be' has temporal reference, then there will be a time when there is time.[368] And what is the difference between these 'times', I mean the 'will be' and the 'when there will be'? And it is just as possible to say the same about the time in which there will be time. And in this way we shall go on searching out more times of times *ad infinitum*. But it is impossible for there to be two times simultaneously and at the same time, ([as there would have to be] for us to concede an infinite number) unless one were part of the other and embraced by it, as if one were to say that a month, for instance, and its tenth day are present at the same time. But the affirmation that 'there will be time' means not this but simply that time will not cease now but will [continue to] be hereafter.

So if, on Proclus' reasoning, an absurdity[369] has followed from the affirmation 'there will be a time', namely, that there is a 'time of time', or that time is in time, then, if there is every necessity that of a pair of contradictory propositions either the affirmative one or the negative one must be true, it must be false and its negation true. So if, as far as Proclus' reasoning goes, the affirmation 'there will be a time' is false, its negation is true; therefore there will not be a time. And consequently the argument has been turned on its head by means of this [same] marvelous mode of argument.

And, since it is also possible to state that there will be a time when there will be rain, say, or anything else that can happen, if here again 113,1 not only the first but also the second 'will be' is – on the assumption that 'will be' is always indicative of time – interpreted temporally, not to mention the time word 'when', how, when we come out with 'time' four times in a row, could the iteration be other than laughable and silly? And it is certainly possible by means of [appropriate] examples to lead the philosopher's reasoning into innumerable other such absurdities.

If, then, 'was' or 'will be' do not in themselves indicate time in the above [examples], but in saying 'there will be a time' [the speaker] is only indicating the future existence of time, and in 'there will be a time, when there will be rain' the first 'will be' again conveys the existence of time, the second that of the rain, then we shall not, if someone says 'there will be a "when" there will not be time', take 'will be' as indicative of time but [only] that there will be a state of affairs in which there will be no time.

And not only is it the case that 'was' and 'will be' are indicative of segments of time,[370] the one of the past, the other of the future, but 'is' itself refers to the present segment of time, which is contrasted with both the past and the future. And when we say 'God is' we certainly do not intend 'is' to refer to a segment of time but only to God's eternal existence. For it is not necessarily the case that when there is inadequacy in the words the subject matter is inadequate too. We should accept from [the words] whatever is useful for the expression of the thought, and reject anything foreign to the nature of the thought that is expressed incidentally.

114,1 In all of the above examples, then, the time words are not, in the context, [to be] taken at all costs as by themselves indicating a time, even if the parts of time *are* concomitantly signified by them. Instead, as we have shown by means of the above examples, they indicate either the existence of something or the end or beginning of some [period] of time. And in the same way we shall not, if it is said that time at some time was not, or at some time will not be, be forced because of the time expression 'at some time' to say either that there was time before time or will be after time. Instead we say that time's being said not to have been at some time is indicative of its having a beginning and not being eternal, and that [its being said] to be going not to exist at some time [is indicative] of its ceasing and not being endless.

But this present argument of mine will not concern itself with whether time will cease or not; it will be enough if what has been said demonstrates that the present proof of Proclus not only lacks cogency but even degenerates into absurdity.

4. One can also say this against the present [proof].

If eternity is the measure of the existence of everlasting things, as time is that of the movement of the heaven, as Proclus himself says in these words: 'time is the measure of the movement of the heaven, as eternity is of the life of the living-creature-itself' – if, as I say, eternity is the measure of the life of the living-creature-itself, that is to say, of everlasting things, there is, one supposes, every necessity that eternity should not be a single point, but a kind of plane or extension, if I may put it so, which is coextensive with the being of eternal things. It is not, like time, cut[371] into various segments, into years, months, days and nights, I mean. Rather, it is as though one were in imagination to stop the heaven and the movement of the sun. The various segments of time would no longer be part of one's mental picture, but there would still nevertheless be a kind of uniform extension keeping pace with the existence of the world. In the same way, I suppose, one could say in regard to everlasting things that, even though for eternity, which measures their being, there is no movement of temporal duration, nevertheless it is certainly the case that a kind of self-uniform extension is thought of in connection with their being. For, as I have said, eternity is not a kind of point without parts and eternal things do not exist just at some one point.

It was for this reason that Proclus, and Plato before him, said that eternity is the measure of the life of everlasting things; for [Plato] says:[372]

> Time came to be along with the heaven, in order that, having been generated together, they may also be dissolved together, should their dissolution ever come to pass. And [it was made] after the pattern of the eternal nature,[373] so that [the heaven] might be as similar as possible to [the pattern]. For the pattern is for all eternity, while [the heaven is] throughout the whole of time.

And so, if Plato says that the pattern for the world has being for all eternity and that eternity is the measure of the life of everlasting things,[374] he clearly does not believe that eternity is a kind of single point (for 'all' is not a point and a measure is not without parts), but [rather that it is] a kind of single and uniform extension of the life of everlasting things, not [internally] divided by any differentiation, but always staying the same and remaining self-identical.[375]

This being so, when we say ‘there was a “when” time did not exist, and there will be a “when” it will not exist’, we do not intend ‘was’ and ‘will be’ temporally. It is the same as when we say ‘God was even before men knew it’ and ‘time was’[376] and ‘God is’ and ‘time is’, or when Plato says[377] ‘He was good’. We are not [in such cases] referring to segments of time by using ‘was’ and ‘is’. On the contrary, we claim that these time words, given the nature of the subject, are indicative of the eternity, or the simple, unadorned existence of the things [they refer to], since we cannot say anything without some reference to time. In just the same way [then], when we say ‘there was a “when” time did not exist, and there will be a “when” it will not exist’, we do not intend ‘was’ and ‘will be’ temporally,[378] but claim that they point to the existence of eternity. For eternity existed even when there was no time, and eternity will exist when there is no time.

And if someone uses the words ‘some time’ in [the sentence] so that it reads ‘there was some time when time did not exist, and there will be some time when time does not exist’, we shall in that case too understand ‘some time’ as referring to eternity, because there is a sort of existence or extension, or whatever else one would like to call it, of eternity in which there was no time and there will be eternity when there is again no time.

As I said[379] [earlier], a sensible person should focus on the intention of a statement and not on any inadequacy in its expression. In our thought about the divine, for example, our understanding does not have the strength to contemplate it without mental imagery. As Plato also says, imagination always goes hand in hand with our conceptions 117,1 about God, assuming on our behalf that it apprehends shape and mass in connection with him. And nor can we think about things that transcend time timelessly. But this weakness of our contemplative faculty does not necessarily mean that things are [really] like this. [So, to avoid any such idea,] we keep at bay any inappropriate features that [threaten to] find their way into our conceptions about God and incorporeal things by applying a mental test. And the same applies, indeed more so, to verbal presentations of a subject: if words cannot always express our thoughts or the subject matter perfectly, we should not focus on the weakness of the expression and base our criticism on the wording rather than the intention of the speaker. [The question at issue is a case in point,] for our argument has shown in a number of ways that time is not in every case part of what is indicated by time words.

And so, in summary, it is rightly held by Plato and the truth[380] that time came into existence along with the heaven (for time is the measure of heavenly movement), and nobody will be able to show that Plato anywhere says[381] that time does not have a beginning; it is on his own initiative that Proclus has been attempting to prove [that it does not].

That time does indeed have a beginning I shall show elsewhere.[382] And Plato too clearly states[383] that it came to be along with the heaven and did not exist before the heaven came to be and says that the heaven came into existence and had a beginning, as Aristotle, Plato's pupil, attests in the eighth book of the *Physics*,[384] confirming that this is the opinion of his teacher. Here is what he says:

> But as far as time is concerned, all with one exception[385] are in clear agreement. They hold that it is ungenerated. Indeed, it is 118,1
> on this account that Democritus shows that it is impossible for all things to have come to be. Plato alone generates it. He holds that it came to be along with the heaven and that the heaven [itself] came to be.[386]

That this is true will be shown by many proofs in chapter six,[387] and it has [already] been shown in [chapter] one that it is impossible for the existence of the world to be without a beginning. And if one or the other of them, whether time or the world, had a beginning to its existence, the other must have had a beginning to its existence as well. And even were one to concede that it is still unclear whether the world and time have a beginning to their existence or not, our argument has adequately demonstrated that the philosopher's present proof at least, by which he hopes to prove that time is without a beginning or an end, is more appropriate to a sophist than to a philosopher.

5. And if he says that eternity is the pattern for time and that therefore there must[388] always be time as well in order that eternity may everlastingly be a pattern, we adequately refuted that kind[389] of argument in the philosopher's second proof, in which he raised the question of patterns. For if, just as living-creature-itself is the pattern for the world, eternity, which is the measure of the life of living-creature-itself, is the pattern for time, which is the measure of the existence of the world, and if it has been shown that living-creature- 119,1
itself does not have its being in being the pattern for the world, then eternity, which is the measure of the life of living-creature-itself, will not have its being in being the pattern for time. For if time is related to the world as eternity is to living-creature-itself, then by permutation eternity will be related to time as living-creature-itself is to the world. But we showed in that [earlier discussion] that just because living-creature-itself, or, more generally, a world-pattern, exists, there is no necessity that the world should also exist. And therefore just because eternity exists, there is no absolute necessity that time should also exist.

The End of the Refutation of the Fifth Argument

Notes

1. (1) The title, headings and section numbers printed in angle brackets at the beginning of the Greek text are Rabe's restorations. The title has been discussed in the introduction. There is no direct evidence that *Aet.* had a preface, but we know that it once had an epilogue and that there was room for one in the missing part of the main manuscript. (According to Rabe, this was equivalent to about forty pages of Greek in his edition, and Proclus' argument and the missing part of Philoponus' reply are unlikely to have taken up any more than twenty. On the other hand, some of the missing leaves may have been blank, and we cannot be sure that the manuscript did not contain anything other than *Aet.*). The other restorations are based on the formatting of the remaining chapters and are certain.

(2) Perhaps this is the best place to say something about the terminology used to describe the various divisions of *Aet.* In the preserved headings for chapters 2-18, which may or may not be Philoponus' own, Proclus' arguments are *logoi*, Philoponus' replies are *luseis* and the sections into which they are divided are *kephalaia*. In the work itself Philoponus' terminology is less consistent. An argument of Proclus is frequently an *epikheirêma* (26,20, etc.), a *logos* may be either an argument of Proclus (126,23, etc.) or Philoponus' reply (69,5, etc.), and a *kephalaion* may be an argument of Proclus (94,22, etc.), a reply of Philoponus (70,9, etc.), or a section of the last (130,11, etc.), in which case it is always a *kephalaion* of a *logos*. In headings I translate *logos* 'argument', *kephalaion* 'section' (a good case could be made for translating *kephalaia* 'Summaries of the Main Points' in these headings – see H.D. Saffrey and L.G. Westerink, Proclus, *Théologie platonicienne*, vol. 1 (Paris, 1968), Notes complémentaires, p. 1, n. 2 – but I have been influenced by Philoponus' usage in the text, which I have outlined above) and *lusis* 'refutation'. In the work itself I translate *epikheirêma* 'proof', *logos* 'argument' when it refers to one of Proclus' arguments, but 'chapter' when it refers to one of Philoponus' replies, and use 'proof', 'chapter' or 'section' for *kephalaion* depending on whether the reference is to an argument of Proclus, a reply of Philoponus or a section of such a reply.

2. As mentioned in the Introduction, the translation of an Arabic version of Proclus' first argument which follows was produced for this series by Peter Adamson.

3. The version of Proclus' first argument translated here is that of Ishâq ibn Hunayn. The text used is A. Badawî, *Neoplatonici apud arabes* (Cairo, 1955), 34.4-35.8. For Jon McGinnis' very helpful translation of the argument, with a modified Arabic text based on Badawî, see Proclus, *On the Eternity of the World*, translated with commentary by H.S. Lang and A.D. Macro (Berkeley, 2001). I have made use of McGinnis' work in revising my own translation. For a French translation, see G.C. Anawati, 'Un fragment perdu du *de Aeternitate Mundi* de Proclus', in *Mélanges de Philosophie Grecque offerts à Mgr. Diès* (Paris, 1956), 23-5. For a German translation by P. Heine, see M. Baltes, *Die Weltentstehung des platonischen Timaios nach den antiken Interpreten*, vol. II (Leiden, 1976), 134-6.

4. The term used throughout for 'everlasting' is *abadî*; the adverb *abadan* (from the same root) I translate as 'everlastingly'. It is at first used as a synonym for 'eternal' (*azalî*). But at the end of the argument there is a distinction between two senses of 'everlasting': God is everlasting in the sense of being outside of time, the world everlasting as existing at all times. God's everlastingness is there defined as *dahr* and *azaliyya*, which I translate on the basis of the context as 'timelessness' and 'eternity'. As McGinnis points out these two terms may correspond to *ho aiôn* and *aïdios*, whereas *abadî* probably translates *aei*.

5. Literally 'the all', *al-kull*, which could translate *to pan*.

6. McGinnis may well be right to suspect that the Arabic words *jûd* and *jawâd*, which I translate 'generosity' and 'generous' (their normal English equivalents), render the Greek words *agathotês* and *agathos*, the normal English equivalents of which would be 'goodness' and 'good'. (To the passages he cites add Proclus, *in Tim.*, 1.366,20-368,11, which closely parallels Proclus' present argument and where *agathotês* and *agathos* are used throughout, and a number of passages in Philoponus' refutation (*Aet.* 7,18-24; 8,20-27; 13,26-14,6) in which God's creative activity is related to his goodness). However, they could also render *aphthonia* and *aphthonos* (more literally 'lack of envy' and 'without envy' or 'ungrudging') and there are in fact two passages in *Aet.* in which God's creative activity is directly related to his lack of envy, or 'generosity'. At *Aet* 13,12 (again in his refutation of the present argument) Philoponus writes: 'If God is good, and, as Plato in his wisdom says, "in the good no envy (*phthonos*) ever arises in regard to anything" [*Tim.* 29E], and [it is] for this reason he brought all things into being, why did he not bring any of [his] creations into existence in all respects identical to himself so that they shared in the same substance and power with him? For either he could have, but did not wish to – and I shall not mention what would follow because it would be blasphemous (for he will not have escaped envy (*phthonos*)) – or it must be that, because he could not, he did not even wish to; but that would be to predicate weakness of God', and at *Aet.* 224,25, paraphrasing a passage from a lost work of Proclus, he writes: 'Plato says "for he was good, and in the good no envy (*phthonos*) ever arises in regard to anything" [*Tim.* 29E], from which Proclus concludes ... that if God is always free of envy (*aphthonos*), [then], since he is also always good, the world too will always exist'. This second passage in particular shows that, despite the evidence of his commentary on the *Timaeus*, we cannot rule out the possibility that Proclus did in fact relate God's creative activity to his 'generosity' and not just directly to his goodness in the present argument.

7. *Kawn*, which is the same word translated as 'coming-to-be' in the previous sentence.

8. *Musâwiyyan*: perhaps meaning 'equal in duration'.

9. Reading *yuhibbu* with Badawî and McGinnis, and rejecting Anawati's emendation to *yajibu*.

10. This interjection, as Anawati rightly says, is surely not based on the Greek.

11. Note that here the word *kawn* (in the form *mukawwin*) needs to refer to generation *in time*, not just bestowal of existence generally, both because of the context generally and because it is opposed to *fasâd*, 'corruption' or 'perishing'.

12. *quwwa*: this word can also mean 'potentiality'. Because of this double meaning *quwwa* is a closer synonym for *dunamis* than *qudra*, which has been translated above as 'ability'.

13. Up until now the verb 'create' has translated *khalaqa* and the noun 'Creator' has translated *al-bârî'*. In this case the noun 'Creator' is *khâliq*. But this, I take it, is synonymous with *al-bârî'*.

14. *Musâwiq*, which means 'going along with' or 'continuous with', 'accompanying'. This may translate the Greek *suzugos* (cf. Proclus *in Tim.* 1.281,3 and 1.295,11).

15. The Greek text as we have it begins part way through a sentence some way into the second section of Philoponus' refutation of Proclus' first argument. To judge from what follows, the missing portion of Philoponus' argument probably went something like this. 'Contrary to what Proclus says, a once-off creation of the world is compatible with eternal and unvarying goodness and power in the creator. A once-off creation would in effect be one aspect of God's failure to create 'more or better' things than he actually has. If it can be shown that this failure in no way detracts from God's goodness or power, the same will automatically follow for a once-off creation. [The advantage of this line of argument is that the issue of God's failure to create 'more or better' thing arises for a Neoplatonist as much as for a Christian]. Now, if 'more or better' things are not created, this is due either (1) to the productive cause (God), or (2) to that which comes into being (the world), or (3) to both [cf. 7,4-12]. If it is wholly (1) or partly (3) due to God, this implies a limitation to his goodness, which would be intolerable [although there is no sign in what remains of the argument that Philoponus dealt with this possibility, I assume that he must have at least mentioned it, if only to dismiss it] or to his power or powers. The possibilities in regard to God's power(s) are that it, or they, is, or are (1) one, simple and finite (2) many and each finite (3) one and infinite (4) many and each infinite, and (5) many and some infinite, some finite [cf. 4,7-24]. Now, all agree that God is eternal, and as Aristotle argues, with, as the following passage shows, the approval of Proclus, that which is eternal has infinite power ...'. Given that the first sections of Philoponus' refutations are normally short, averaging just over two pages in Rabe's edition, with only one (8.1 at 7 1/2 pages) much over three, it is quite likely that this material took up most, if not all, of the missing part of the first refutation.

16. Rabe suggests that Philoponus is quoting Proclus, who is in turn referring to Aristotle, and in particular to his remarks on time and eternity at *Cael.* 279a11-b3, and punctuates accordingly. He may be right, but it seems more likely that Proclus is the subject of *legei* ('he says'), that the reference is to the last few sentences of the first argument (which Rabe did not have) and that Aristotle was cited earlier, so I have omitted Rabe's closing quotation mark.

17. For a discussion of the role of 'infinite power' arguments in the creation debate in ancient and mediaeval philosophy with references to earlier literature, see R. Sorabji, *Matter, Space and Motion* (London & Ithaca, NY, 1988), ch. 15. For Philoponus' contribution in *Aet.* and elsewhere, see especially pp. 254-9. (The same material appears in abbreviated form in R. Sorabji (ed.), *Aristotle Transformed* (London & Ithaca, NY, 1990), ch. 9).

18. i.e. everything there is must be either eternal or temporal. (The clause takes the form of a step in a formal division by dichotomy.)

19. Accepting Rabe's suggested restoration of the (only partially legible) Greek text at 2,8.

20. The distinction between *aiônios* ('eternal') and *aïdios* ('everlasting') is discussed in the Introduction.

21. i.e. not complex.

22. In most contexts *hama noêmati*, which also occurs at 18,33 and 149,9, is merely a colourful way of saying 'very quickly' or 'instantaneously' ('quick as thought'), but the reference to creation by thought at lines 10-11 below suggests that it may have further connotations here and I have translated accordingly.

23. *Psalms* 33,9.

24. *Judith* 9,5-6.

25. For Proclus too, of course, creation results from intellection; see e.g. *ET* 174.

26. Changing *paragesthai* to *paragetai* at 6,14, as suggested by Rabe in the critical apparatus.

27. The *Ten Puzzles*, is one of three short, closely-connected essays on providence by Proclus which survive only in thirteenth-century Latin versions by William of Moerbeke. The present passage (5,32-5 in the edition of Boese, 5,43-8 in that of Isaac), one of three, all from the same chapter near the beginning of the work, which Philoponus makes use of in *Aet.*, is also quoted at 38,16-20; 91,19-23 and 570,14-18.

28. i.e. physical rather than metaphysical.

29. *Phys.* 204a8-206a8.

30. *Cael.* 271b1-276a17. (In the translation I translate all titles into English, but in the notes I normally use the traditional abbreviations, which are frequently formed from the Latin versions of titles.)

31. Aristotle's arguments against the possibility of an actual or traversed infinity and Philoponus' use of them against Aristotle himself and the Neoplatonists in *Aet.* and elsewhere are discussed by Sorabji in *Time, Creation and the Continuum* (London & Ithaca, NY, 1983), ch. 14, and in *Philoponus and the Rejection of Aristotelian Science* (London & Ithaca, NY, 1987), ch. 9.

32. This last statement is clearly false as it stands, but perhaps we should regard it as shorthand for something like 'for that which consists of a finite number of finite parts is finite'. (Philoponus seems to have been influenced by the phrasing of the similar argument at *Cael.* 271b19-23 where Aristotle initially says that if the parts of a composite body are finite, it too must be finite, even though he immediately makes it clear that this means 'finite in both number and magnitude'.)

33. Other possible references to this work occur at 11,16; 117,20-1; 259,3-5 and at *in Phys.* 430,9-10. (References from Verrycken, 'The Development of Philoponus' Thought and Its Chronology' in R. Sorajbi (ed.), *Aristotle Transformed* (London & Ithaca, NY, 1990), 254. Incidentally, given the similarities between their respective contexts, Verrycken seems needlessly tentative when suggesting that the *in Phys.* reference and the present one refer to the same work). Although it has not survived in its own right, it may be the work known through an Arabic summary translated into English by Pines (S. Pines, 'An Arabic Summary of a Lost Work of John Philoponus', *Israel Oriental Studies* 2, 1972, 320-52) and, according to Wildberg, possibly also the work attacked by Simplicius at *in Phys.* 1326,38-1336,34. (See C. Wildberg, *Simplicius: Against Philoponus on the Eternity of the World* (London & Ithaca, NY, 1991), 100; but Wildberg also cites evidence from Arabic bibliographies which suggests that this may be yet another work on the perishability of the world).

34. i.e. after completing *Aet.* and probably, as Verrycken ('The Development of Philoponus' Thought and Its Chronology', 251) argues, the *contra Aristotelem* as well.

35. The verb (*exarithmeisthai*) is presumably middle, which is unusual in this sense.

36. Inserting *legô dê* after *adunaton* at 10,19, one of two possible corrections suggested by Rabe in the critical apparatus. (Another possibility is that the words *to hoionei aparithmoumenon kata monada diexelthein to apeiron* are a marginal gloss that has entered the text).

37. Adding *anthrôpôn* before *tôn* at 11,3, as suggested by Rabe in the critical apparatus. (The case for the addition is improved by the occurrence of the same phrase in a closely related passage at *in Phys.* 428,25-6.)

38. Adding *ei* after *kai* at 11,6.

39. See the note on 9,22 above.

40. 9,14-11,7 are translated in R. Sorabji, *Time, Creation and the Continuum*, 214-15.

41. More literally 'runs alongside the thought of [our] mind'.

42. See the note at 7,25.

43. sc. that of mules.

44. *GC* 336b31-2.

45. *Tim.* 29E.

46. *ousian*. In this section, as often, the choice between 'being', 'substance' and 'essence' to render *ousia* is not clear-cut. The linking of *ousia* with *dunamis* ('power') here and with *dunamis* and *energeia* ('activity') later in the section (14,16-17) perhaps suggests 'essence', but the fact that the *ousia* under discussion might, hypothetically at least, be created (13,24-25), suggests that 'being' or 'substance' might be more appropriate. A case could be made for either, but I have opted for 'substance', largely because I need 'being' for *to einai* when the phrase *to einai kai tên ousian* occurs at 14,11. (Incidentally, later in *Aet.* I several times translate this same phrase 'being and essence'.)

47. Literally 'that substance', or 'substance there'. *ekeinos* and *ekei* are frequently used to contrast things in the divine or intelligible sphere with things here.

48. Because I reserve 'eternal' for *aiônios* and translate *aïdios* 'everlasting', I have coined the unlovely word 'co-everlasting' to render *sunaïdios* rather than use the more usual 'coeternal'.

49. Changing *proüparkhon* to *proüparkhoi* at 14,20, one of two possible corrections suggested by Rabe in the critical apparatus.

50. For Neoplatonists from Plotinus onwards the three hypostases One, Intellect and Soul exercise causation while themselves remaining absolutely immutable. They give rise to the things that participate in them without deliberation or action on their part just by being what they are. The present phrase (*autôi tôi einai*) was probably first used to express this mode of causation by Syrianus and became a stock formula with later Neoplatonists, being a favourite with Proclus, who uses it in the second, fourth and sixteenth arguments (see 24,5; 56,2; 560,22; 561,25.26), and occurring 27 times in *Aet.*, mostly in Ch. 7, where the manner in which soul initiates life and movement is at issue. It is to the extent that the sun can be said to produce light 'just by being' that it provides a suitable analogy for the manner in which the creator produces the world. (For the Plotinian background and the meaning and employment of *autôi tôi einai* see C. D'Ancona Costa, 'Plotinus and later Platonic philosophers on the causality of the First Principle' in L.P. Gerson (ed.), *The Cambridge Companion to Plotinus* (Cambridge, 1996), 356-68).

51. For the parallel between the propagation of light and creation, see e.g. Plotinus, 5.1.6,25-39; Proclus, *in Tim.*, 1.290,29-291,1. For Proclus all higher-level causes produce their effects 'just by being'. (See e.g. J. De Groot, *Aristotle and Philoponus on Light* (New York & London, 1991), 84 ff. with earlier literature cited there, and R.J. Hankinson, *Cause and Explanation in Ancient Greek Thought* (Oxford, 1998), 415 ff. for the antecedents of the doctrine in Plotinus and others).

52. Evidently in the missing part of the present chapter.

53. Astronomy and geometry perhaps, or even optics.

54. *DA* 418b9-11.

55. Rabe is of the opinion that the words I translate 'Hence there has to be a period, during which [the sun's] rays do not reach under [things]' are corrupt, and it does seem possible that the words 'during which [the sun's] rays do not reach under [things]' (*en hôi hai aktines oukh hupoballousin*) are an explanatory gloss on 'a time' (*khronou tinos*) which has found its way into the text. However, if we assume that *hupoballein* is being used intransitively (although LSJ gives no instances of the intransitive use of *hupoballein* there are instances where *epiballein* is so used in similar contexts) and that 'reaching under' includes striking objects from an angle rather than from directly above, it seems possible to retain the text of the manuscripts.

56. Changing *ekeinôn* to *ekeinou* at 16,18, as suggested by Rabe in the critical apparatus.

57. In what follows Philoponus tries to show that an analogy between the propagation of light and the generation of the universe fails to support the Neoplatonic view of generation on *any* view of the nature and propagation of light. This being so, some of his arguments are intended to be perfectly general, some are directed specifically at Proclus or other unnamed opponents, and some presuppose his own position on the propagation of light, which he does not reveal until near the end of the chapter, at 21,24 ff.

58. As Rabe suggests ('haec vix sana'), the clause beginning *epeidê to phôs* (16,25 ff.) is unsatisfactory as it stands. I suspect that the phrase *epeidê to phôs ditton estin* ('since the light is twofold') was originally a gloss on *to men phôs* explaining the occurrence of *men* ('on the one hand'), which is not taken up by its *de* ('on the other hand') until the first line of the following section, and that the word *sunupostan* ('having come into existence simultaneously with it') was originally a gloss on *ousiômenon* and have accordingly omitted both in the translation.

59. More literally, 'foreign and alien'. (See the entries in Lampe for similar usages of the two words).

60. *en tôi ginesthai to einai ekhei.* The phrase, which is common in Philoponus and Simplicius and is used by a number of the other commentators, goes back at least to Alexander (*Mixt.* 227,22; *in Sens.* 50,25; 154,16; *in Meteor.* 73,1; 83,1). At *in Phys.* 465,25 Philoponus uses it to distinguish processes or events, such as a day or a contest, which unfold little by little and are never present as a whole, from substances, such as a man or a horse. However, in the Platonic tradition substances, and indeed the world as a whole (see *Tim.* 27E ff.), are regarded as ever-changing entities and in *Aet.* the phrase is most commonly used to distinguish the physical from the intelligible and the temporal from the eternal.

61. Rather more literal renderings would be: 'the illuminating [body]', or 'that which illuminates'.

62. *Tim.* 31B.

63. The word *energeia* may be used to describe (1) the process whereby a form is realised (= 'actualisation'). (2) the manifestation of the realised form – where it is a virtual synonym of *entelekheia* – (= 'actuality') (3) the active putting forth of power (= 'activity'). In practice these senses often overlap and it can be hard to settle on a translation. In what follows it may be that 'actuality' best describes the light 'in' the sun, 'activity' the light radiated by the sun as Plotinus understands it, and 'actualisation' light present in a transparent medium as Aristotle and Plotinus himself understand it, but I have thought it best on balance to keep to the single rendering 'activity'.

64. For the phrase see the note at 6,3.

65. Or perhaps 'either implicated in matter or free of matter', but *ta enula* at 18,22 seems to mean 'material objects'.

66. Changing *hupo* to *epi* at 18,26, as suggested by Rabe in the critical apparatus. (Proclus' *On Light* has not survived.)

67. Or perhaps 'in [their] form'.

68. For a brief account of Proclus' rather complex metaphysics and physics of light, see L. Siorvanes, *Proclus: Neo-Platonic Philosophy and Science* (New Haven & London, 1996), 241-4.

69. There does not seem to be anything to this effect in Plato, although he may have in mind *Rep.* 508E-509A, where Socrates says that it is right to think of light and vision as *hêlioeidê*, or 'sun-like'. (Did he perhaps read *hêliou eidê*, or 'forms of the sun'?).

70. More literally 'will fall even further from the matter under investigation'.

71. cf. *Enn.* 4.5.6-7. (Neither of the statements is a genuine quotation).

72. sc. the actuality of light.

73. Plotinus himself would not accept this conclusion. Although the following passage uses the example of fire and heat, it illustrates his actual position particularly well. 'In each and every thing there is an activity (*energeia*) which belongs to substance and one which goes out from substance; and that which belongs to substance is the active actuality (*energeia*) which is each particular thing, and the other activity derives from that first one, and must in everything be a consequence of it, different from the thing itself: as in fire there is a heat which is the content of its substance, and another which comes into being from that primary heat when fire exercises the activity which is native to its substance in abiding unchanged as fire. So it is also in the higher world ...'. (*Enn.* 5.4.2,27-34, tr. Armstrong). It is because radiated light is *not* constitutive of a light-source for Plotinus that he can use it to illustrate the procession of everything else from the One without any effect on it. (For light as a metaphor for procession in Plotinus, see F.M. Schroeder, 'Plotinus and language', in *The Cambridge Companion to Plotinus*, ed. L.P. Gerson (Cambridge, 1996), 341-3, who quotes the above passage on p. 340).

74. Philoponus sometimes puts verbs introducing quotation or report in the present tense and sometimes (as here) in a past tense. In either case I normally use the present tense in the translation.

75. *DA* 418b9-14. In fact Philoponus' own views on the nature and propagation of light are developed in the course of commenting on *DA* 2.7 (from which the present passage is quoted) and are in essential agreement with those of Aristotle. There is a detailed account of the key passage from *in DA* (324,23-342,16) and of relevant passages in other commentaries in J. De Groot, *Aristotle and Philoponus on Light*; later developments in Philoponus' theory of light in *Opif.* can be found in C. Scholten, *Antike Naturphilosophie und christliche Kosmologie in der Schrift 'De Opificio Mundi' des Johannes Philoponus* (Berlin & New York, 1996), 235-70; there is an assessment of Philoponus' originality vis à vis Aristotle in R. Sorabji, *Philoponus and the Rejection of Aristotelian Science* (London & Ithaca, NY, 1987), 26-30.

76. i.e. aether, the material of the spheres that carry the heavenly bodies.

77. Aristotle has 'for to this too belongs something that is one and the same' (*kai gar toutôi ti huparkhei hen kai tauton*), which must mean that aether has something in common with fire – something which accounts for the fact that it too can illuminate the transparent. In Philoponus 'this' presumably refers to something like 'the ability to activate the transparent' and 'these' to fire and other things with similar properties, including aether.

78. Literally 'second'.

79. Proclus had the title Successor (*Diadokhos*) as head of the Academy at Athens in the line of succession from Plato.

80. A closer rendering would be 'and this is its essence, being a paradigm'. Similar phrases occur later in the chapter at 25,8; 26,8; 32,22; 33,25; 34,4.11, where I also telescope the construction. On the phrase *to ti ên einai* and the possessive (?) datives that frequently accompany it (in this case *autôi* and *paradeigmati*), both of which originated with Aristotle, see J. Owens, *The Doctrine of Being in the Aristotelian 'Metaphysics'* (Toronto, 1963), 180-7. (Owens does not approve of the translation 'essence'.)

81. Or perhaps 'property', or 'function' (see LSJ *dunamis* II.b).

82. *autôi tôi einai*. For the implications of the phrase, see the note at 14,25.

83. Something is clearly wrong with this sentence and Rabe writes: '9 sq.: haec vix sana'. On balance it seems to me most likely that a phrase referring to the *paradeigma* has dropped out after *hote ouk ên* at l. 10 and I have added the words *kai to paradeigma ouk ên* at that point and deleted *kai to paradeigma* at 24,11 on the hypothesis that the words *kai to paradeigma* became displaced at some point in the transmission of the text and that *ouk ên* was then deleted as a dittograph. (Another possibility is that *kai to paradeigma* was accidentally omitted because of the repetition of the same phrase in the next line and *ouk ên* again deleted as a dittograph. In that case I would translate the last part of the sentence: 'unless the very pattern, there being no copy, is not a pattern at all, or is not the pattern for the copy [in question].' – Baltes would simply add *aei estin* after *eikôn* at 24,9.)

84. Translations into modern languages of the seventeen arguments of Proclus which survive in *Aet.* are listed in the Introduction.

In the notes to this translation I normally do no more than identify direct quotations and obvious references, often merely repeating Rabe's identifications from his critical apparatus. In the case of Proclus' arguments, a wider range of comparative material can be found in the notes to H.S. Lang and A.D. Macro (eds.), *On the Eternity of the World* (De Aeternitate Mundi), *Proclus* (Berkeley, etc., 2001).

85. For *epikheirêma*, LSJ gives: [I] *undertaking, attempt* ... II. in the Logic of Arist., *attempted*, i.e. *dialectical, proof* ... (at *Top.* 162a16 Aristotle himself defines it as 'a dialectical deduction'). This being so, one might be tempted to think that Philoponus' application of it to Proclus' arguments is tendentious, but in the commentaries he routinely describes Aristotle's arguments as *epikheirêmata* and it seems probable that he is simply using it as a neutral word for a proof or line of argument. (At *Aet.* 406,7 and 19 it is actually used of his own arguments, though admittedly in a contents summary).

86. More literally 'has its being in this, in being ...'. Philoponus uses expressions of this form very frequently both in *Aet.* and elsewhere when stating a thing's essence. Outside Philoponus they are not very common (a TLG search turns up four in Alexander, two in Porphyry, eight in Simplicius – one of them in a quotation from Philoponus – and a handful in other later commentators), and I cannot account for the mannerism. I have not made any attempt to preserve the construction in translation.

87. Proclus does not explicitly claim that either hypothesis is Platonic or to be found in Plato. He would doubtless have held, as Philoponus himself does, that the first is, citing the *Timaeus*, but might not have claimed Platonic warrant for the second, which is expressed in Aristotelian language. Later, in the argument itself (at 26,14-18 and 33,3-4), Philoponus writes as though it was, or would have been, Proclus' position that the first premiss implies the second and that, since the first is in Plato, he would have had to accept the second. Both here and there Philoponus seems to be putting words into Proclus' mouth. Perhaps there was something in the pages missing from the beginning of *Aet.* to explain why he writes in this way.

88. cf. *Tim.* 28A-29A.

89. This principle of the simultaneity of relatives was first enunciated by Aristotle (*Cat.* 7b15-22), although he immediately (7b22-8a12) cast doubt on its universality.

90. As Philoponus well knew, the orthodox later Neoplatonic position was that there was in reality no such disagreement. (See the note at 29,6 for further detail).

91. The precise reference of *to toiouton* is unclear. Initially it is natural to take it as referring to Socrates' interest in ethical questions and neglect of the world of nature, but the final clause of the sentence suggests that Aristotle has in mind the Heraclitean opinions that Plato absorbed from Cratylus. Perhaps the truth is that, despite being in the singular, it encompasses both. In any case, it should refer back to something that has gone before and not forward to the next clause, as Ross, the Oxford translator, seems to believe. (Philoponus' paraphrase at 28,9-18 perhaps suggests that he took *to toiouton* to refer to Socrates' methods and the final clause of the sentence to Plato's Heracliteanism.)

92. sc. the search for the universal and the construction of definitions.

93. Omitting *aei*, which does not occur in Aristotle and, in view of the occurrence of *tinos aei* in the next line, looks suspiciously like a case of dittography.

94. *Metaph.* 987a29-b10.

95. *Tim.* 28A.

96. Changing *genomena* to *ginomena* at 28,18, as suggested by Rabe in the critical apparatus.

97. Omitting *einai* at 28,23. (Rabe writes '*einai* suspectum'.)

98. From Porphyry onwards most of the Neoplatonic commentators on Aristotle held that, contrary to appearances, Aristotle and Plato were in agreement on most matters. The view that there was no disagreement between them on the theory of Forms can be traced back at least to Iamblichus (Elias *in Cat.* 123,1-3) and was the orthodox Neoplatonic position from the time of Ammonius, who claimed that for both Plato and Aristotle the Forms were creative principles (*dêmiourgikoi logoi*) in the creative Intellect and that Aristotle's criticism of the Forms is not directed at Plato, but at an incorrect interpretation of his theory which gave independent existence to the Forms. This position can be found in Asclepius *in Metaph.* and in the unrevised portions of Philoponus' earlier commentaries, where, certainly in the first case and probably in the second, the teaching of Ammonius is reported, and in Simplicius. The 'more recent [commentators]' (*neoteroi*) referred to are, therefore, probably Ammonius and his pupils and Philoponus is attacking a position that was in some sense at least once his own. This being so, it is perhaps a little surprising that he should accuse those who adopted it of gross effrontery (*huperbolên tês anaideias*) at 32,11. (For further details and references to the commentaries, see the discussions of Sorabji (pp. 2-5) and Verrycken (pp. 215-26) in R. Sorabji (ed), *Aristotle Transformed* and the latter's 'Philoponus' Interpretation of Plato's Cosmogony', *Documenti e Studi Sulla Tradizione Filosofica Medievale* 8 (1997), 270-81; of course the location of the Forms in the mind of God was already Middle Platonist doctrine.)

99. Changing *parakousantas* to *epakousantas* at 29,10, one of two possible corrections suggested by Rabe in the critical apparatus.

100. *DA* 1.

101. *An. Post.* 83a32-3.

102. *Metaph.* 991a20-2.

103. The saying occurs again at 144,21-2. For Philoponus' part in the development of the topos see L. Tarán, 'Amicus Plato sed magis amica veritas, from Plato and Aristotle to Cervantes', *Antike und Abendland* 30 (1984), 112-15. Verrycken also discusses Philoponus' use of it in 'Philoponus' Interpretation of Plato's Cosmogony', 274-7.

104. Verrycken (op. cit., 275) suggests that Philoponus must have seen a parallel between Aristotle's 'delicate position' in relation to Plato and his vis à vis his own teacher Ammonius. This is likely enough, but I cannot agree with him when he goes on to suggest that Philoponus' preoccupation with his own situation has led him to (consciously or unconsciously?) paraphrase Aristotle inaccurately. The issue seems to turn on the meaning of the phrase *ta oikeia anairein* at 31,3-6. Philoponus paraphrases this *tas oikeias* [*hupotheseis elenkhein*], or '[to scrutinise] our own [hypotheses]', which Verrycken rejects in favour of Rackham's 'to sacrifice ... one's closest personal ties' in the Loeb Classical Library translation. But, although Verrycken proceeds as though there can be no doubt that Rackham's interpretation is the 'correct' one, this is by no means obvious; Ostwald, for instance, takes the same view of the passage as Philoponus and translates '[to] give up ... [theories that once were] his own'. This is not the place to consider the pros and cons of the two interpretations. (As it happens, I find that of Philoponus the more plausible). It is enough to point out that the issue is not clear cut and that there is no reason to see Philoponus' interpretation of the phrase in question as evidence of any discomfort or guilt on his part, as Verrycken suggests. In fact, it is quite possible that it did not occur to him that there was any other way of understanding the phrase. (I have deliberately been speaking of interpretation rather than translation because the versions of Rackham and Ostwald are as interpretative as that of Philoponus. The precise reference of *ta oikeia* (more neutral renderings of which would be '[one's] own things', or 'what is [one's] own') is left unclear and a translator probably does best to retain the imprecision of the Greek, as does Ross's 'to destroy what touches us closely' in the *Revised Oxford Translation* – notwithstanding which in the next paragraph I shall translate Aristotle's words 'to demolish even our own [positions]' in line with Philoponus' understanding of their meaning).

105. *EN* 1096a11-17.

106. Adding *kai* before *diapherontôs* and *en* after it at 31,8, as suggested by Rabe in the critical apparatus.

107. This work is not extant. Philoponus quotes from it or refers to it some twenty times in *Aet.*, Proclus himself refers to it in his commentary on Plato's *Timaeus* and Simplicius probably draws on it in a number of his commentaries. The quotations in *Aet.* and some probable quotations in Simplicius' *in Cael.* are translated by Thomas Taylor in *The Fragments That Remain of the Lost Writings of Proclus, Surnamed the Platonic Successor* (London, 1925), 2-31 and the work is discussed in L. Siorvanes, *Proclus: Neo-Platonic Philosophy and Science* (New Haven & London, 1996), 216-23. Philoponus' citations are listed in Rabe's index of proper names under 'Proclus'.

108. The expression *epi lexeôs*, which literally means something like 'verbatim' or 'word for word' and which I here translate 'to quote his exact words', was one method used to mark direct quotations in the absence of typographic indications and a case could be made for not translating it at all.

109. *Metaph.* 991a20-2.

110. The word here translated 'living-creature-itself' (*autozôion*) is frequently used by Proclus in his commentary on the *Timaeus* to refer to the intelligible living creature (*zôion*) of which the universe is a copy in the *Timaeus*. Aristotle does not use the word with this reference and does not discuss the *zôion* of the *Timaeus* in the *Metaphysics*. He does, however, use the words *zôion* and *autoanthrôpos* (man-himself) when criticising the paradigmatic function of the theory of Forms immediately after describing the word 'pattern' as metaphorical (991a28-9), and Proclus may have had this passage in mind.

111. *An. Post.* 83a33.

112. *EN* 1096a11-1097a14.

113. *GC* 335b9-24.

114. Not quite, of course (see the note at 29,6).

115. The promise must have been made in the pages lost at the beginning of *Aet.*

116. Changing *huphestêken* to *sunestêken* at 33,2, as suggested by Rabe in the critical apparatus.

117. The term 'first substance' is actually Aristotelian and Aristotle applies it to individuals in the physical world, but he would have agreed that Plato is treating the Forms as first substances by giving them separate existence.

118. For 'perfect living creature' see *Tim.* 31B and for 'always being the same and unchanging' *Tim.* 29A. 'Living creature itself' (I have assumed that *auto zôion* is equivalent to the more usual *autozôion*), although much used by the Neoplatonists, does not occur anywhere in Plato. The pattern of the world is nowhere said to be 'truly existent', although sensible objects are said *not* to be at *Tim.* 28A and the phrase is applied to Forms in other dialogues.

119. Deleting *allou deomenon eis huparxin* at 33,16 and adding *deomenon eis huparxin* after *allou* at 33,17, as suggested by Rabe in the critical apparatus.

120. The construction is awkward and Rabe may be right to suggest that *hôst'* at 33,23 has replaced *malist'*, which would take up *hêkista* at 33,20. I would then translate: 'So if the pattern of the world is a substance, it above all would have …'.

121. cf. 33,7-8.

122. The reference appears to be to the second 'definition' of relatives at 8a31-2.

123. cf. Aristotle, *Cat.* 10.

124. Philoponus could have invoked the authority of *Cat.* 8a-b24 for this position.

125. 33,26-34,2.

126. *Rep.* 592B.

127. Repositioning *ousias* to follow *legô* (within the parentheses) at 35,28.

128. The words translated 'subject' and 'portrait' in this and the next few sentences are the words I have been translating 'pattern' and 'copy'.

129. Adding *allôn tôn* after *tôn* at 36,14, as suggested by Rabe in the critical apparatus.

130. It is often appropriate to translate *Hellênes* 'Greeks' and in a number of passages in *Aet.* I do, but for Jews and Christians alike, including those of Greek culture like Philoponus himself, the word is often equivalent to 'heathen' or 'pagan', and in contexts where it seems to have that connotation I translate 'Hellenes' rather than 'Greeks'. (For this expedient, cf. L.G. Westerink (ed.), *Anonymous Prolegomena to Platonic Philosophy* (Amsterdam 1962), xiii).

131. The two Greek words (*manteia* and *khrêstêria*) can both be translated 'oracular shrines'.

132. The Olympian gods, to whom these 'shrines and oracles' were normally dedicated, are frequently referred to as *daimones*, or 'demons', by the Christians. For the pagans *daimones* were minor deities or spirits (normally without pejorative overtones), and the Olympians were *theoi*, or 'gods'. The word here translated 'pretended' (*hupokrinesthai*) can also mean 'replied' or 'interpreted' and is one of the words that is used of giving an oracular response; there seems to be a play on words. As I indicated in the introduction, the sentence is a clear echo of the last part of the last sentence of Eusebius, *Praeparatio Evangelica*, 5.3.10. Evil demons appear again in *Aet.* at 241,16; 635,15 and 644,2.

133. 5,1-8 (Boese) (= 5,1-11 (Isaac)).

134. 5,15-25 (Boese) (= 5,21-34 (Isaac)). Also quoted at 91,10-18 (with the omission of 5 lines in the middle) and at 570,1-13.

135. 5,32-5 (Boese) (= 5,43-8 (Isaac)). Also quoted at 6,17-21; 91,19-23 and 570,14-18.

136. Philoponus' language here is obviously more appropriate to the Christian God than to the Neoplatonic One, but although in other works Proclus is usually careful to avoid attributing specific activity of any kind to the One itself, the language he uses in the theodicy of the *Ten Puzzles* does go some way towards justifying it. (Elsewhere, although both extend to all three hypostases, the primary locus of creative activity is in Intellect and that of providence at the level of the divine Henads in the first hypostasis).

137. *Enn.* 6.7.1,45-57. The usual title of the treatise is *How the Multitude of the Forms Came Into Being, and On the Good.*

138. Although Philoponus makes much of this 'on account of' (*dia*) in the explication of this passage which follows, the text of Plotinus actually has *meta* ('after') as it does in the line after next.

139. *hen* ('one') does not occur in the text of Plotinus and is absent both when Philoponus quotes this sentence at 40,22 during his analysis of this passage and on the two other occasions when he quotes this passage (101,24 and 572,7), and I have omitted it from the translation.

140. See the note at 39,15.

141. Changing *touto* to *hoper* at 40,12 rather than adding *auto de* before *touto* at 40,13 as Rabe suggests in the critical apparatus.

142. Omitting *hen*, which Rabe inserts here on the strength of its occurrence in the manuscripts at 39,7.

143. The 'belief in regard to fate' is, as the rest of the sentence makes clear, astrology, or perhaps astrology along with other methods of prognostication. As a Christian, Philoponus could only be hostile to astrology and he attacks it at some length in the later work *Opif.* Here he is going to concede its validity for the sake of the argument, so a direct attack on it would not serve his purposes, but the failure to name it directly, the use of the word *hupolêpsis*, which, like *hupothesis*, often has a pejorative ring when used to describe the beliefs of others (in *Opif.* he refers (p. 195) to the 'absurd notions' (*atopoi hupolêpseis*) of those who believe in astrology and devotes a whole section (pp. 199-204) to arguing that it does not deserve to be called an art), the references to lack of consensus and to 'pagans' and the brevity of the treatment all combine to show his distaste for it.

hupolêpsis, however one translates it (other possible renderings are 'supposition', 'assumption', 'notion', 'prejudice'), makes an awkward subject for the main verb of the sentence. Perhaps Philoponus thought of the phrase as equivalent to 'the pseudo-science astrology', or something similar, or perhaps he just lost the thread, but I would not rule out the possibility that something has gone wrong with the text.

144. Literally 'the children of the Hellenes', but the phrase is merely periphrastic for 'the Hellenes', or 'the pagans'. (For the usage, see LSJ *pais* I.3).

145. According to LSJ, *katarkhê* was a technical astrological term for a forecast concerning an undertaking or voyage. However, the literal meaning of the word is 'starting-point', and the anonymous author of the gloss in the chief manuscript of Philoponus, who suggests that it is here equivalent to 'date of birth', is probably on the right track, in which case the reference is specifically to 'nativities' or horoscopes.

146. In *Opif.* (pp. 195-7) Philoponus reproduces from Origen a proof that even if astrology is able to predict future events it does not bring them about.

147. On balance I think it likely that Proclus is already talking about the creator of the universe and not, as Lang and Macro's translation 'If the demiurge of something is a demiurge' implies, about any creator at all, and have translated accordingly.

148. Rabe believes that Philoponus has *An. Post.* 98a35 ff. in mind and in view of the reference to Alexander of Aphrodisias' commentary on *An. Post.* at 48,10 he may well be right, but *Phys.* 195b16-21; 27-8 is also a possibility.

149. 26E-27A.

150. At e.g. *Metaph.* 1049b24-9.

151. Literally 'former', 'earlier'. The axiom was stated at 42,7-8.

152. Stated at 42,15-17.

153. Adding *ginesthai* after *energeiâi* and retaining the manuscript reading *aitiôi* at 43,14. (Another possibility would be to change *energeiâi* to *energein*, again retaining *aitiôi*).

154. The awkwardness of the construction at this point reflects a corresponding awkwardness in the Greek.

155. Omitting *heteron* at 43,22. (Rabe writes '*heteron* suspectum', although Baltes, who also omits it, reports that the Arabic translator Is<u>h</u>âq ibn <u>H</u>unayn appears to have found it in the Greek).

156. For the translation of *hexis* see the note at 46,12.

157. Inserting *to* after *hoti* at 44,12, as suggested by Rabe in the critical apparatus.

158. *eisagein* often seems to invite this translation in Philoponus and, even though 'God' is a rather awkward subject for 'implies', it seems to work better here than alternatives such as 'introduce', 'import', 'bring [with him]'.

159. Omitting *energeiâi* at 44,21. (Rabe suggests transposing *dêmiourgou* and *energeiâi* but in light of 52,8 it is perhaps more likely that *energeiâi* is a gloss that has found its way into the text).

160. Changing *ton* to *to* at 44,24, as suggested by Rabe in the critical apparatus.

161. Changing *autos* to *auto* at 44,25, as suggested by Rabe in the critical apparatus.

162. For literature on this concept of 'fitness' or 'suitability' see Siorvanes, *Proclus: Neo-Platonic Philosophy and Science* (New Haven & London, 1996), 200, n. 23. (The reference there to pp. 104-9 of Sambursky must be to his earlier discussion in *The Physical World of Late Antiquity* (London, 1962)).

163. Changing *onta* to *gnonta* at 45,12, as suggested by Rabe in the critical apparatus.

164. Adding *tou* before *energeiâi* at 45,16, as suggested by Rabe in the critical apparatus.

165. Adding *tou* before *dunamei* at 45,17 and before *energeiâi* at 45,18, as suggested by Rabe in the critical apparatus.

166. See especially *DA* 417a21-b2, which Philoponus quotes and explicates in IV.7, and on which his own account in this section is largely based.

167. Since for Aristotle *grammatikê* probably meant reading and writing (cf. *Top.* 142b31-3 and his use of the letter A as an example of the sort of thing that a practitioner of the art has knowledge of at *DA* 417a29) and *in DA* 300,3-5 shows that Philoponus appreciated this, a case could be made for translating *grammatikê* 'reading and writing' and *grammatikos* 'literate'. However, in similar contexts in his *Categories* commentary (cf. especially 142,21-6), Philoponus clearly has in mind the academic discipline of 'grammar', which in Philoponus' day embraced philology, literary criticism and much else in addition to grammar, and in *Aet.* too he seems to have in mind something more elaborate than elementary reading and writing, so I have opted for 'grammar' in the case of *grammatikê* and, construing it substantively, for 'grammarian' in the case of *grammatikos*. (Philoponus, who was known as 'the Grammarian', may well have taught 'grammar' himself).

168. The verb translated 'acquiring from others' may, but need not, imply a formal teaching situation.

169. At *in DA* 301,21-3 in a similar context Philoponus uses the example of geometrical theorems and on balance it seems best to use 'theorem' to translate *theôrêma* here in *Aet.* even though the sole example of a grammatical *theôrêma* in *Aet.* is *tode to alpha* ('this particular A') at 71,10-12, which, coupled with the fact that Lampe (s.v. 5) cites a passage in which parts of speech are described as *theôrêmata*, suggests that the *theôrêmata* of grammar are not what we would normally call theorems. (*in Cat.* 193,13-27 perhaps suggests that they would include letters, syllables, words and sentences). Perhaps something like 'the basic elements identified by theory' would catch the sense of the word, but that would be far too cumbersome. (Of the alternatives offered by the lexica '*datum* or *rule of art*' (LSJ II.1.b) seems to come closest, but it does not quite fit the bill).

170. It is difficult to find a satisfactory single rendering for *hexis* when it is used as part of the Neoplatonic vocabulary of potentiality and actuality. Here, and in a number of other passages, I have opted for 'possession' (see LSJ I.1), but more often something in line with LSJ II ('*a being in a certain state, a permanent condition* as produced by practice') seems appropriate. In this vein 'disposition' is attractive, not least because 'dispositional[ly]' would often provide a neat solution for the sometimes difficult phrase *kath' hexin*, but I have opted for 'capacity' (cf. LSJ II.3 '*trained habit, skill*') because all of Philoponus' examples involve the acquisition of a capacity or skill and because it seems to work better in contexts like 48,2-4. Finally, at 69,25 ff., where Philoponus quotes and comments on one of the Aristotelian passages that lies behind the Neoplatonic scheme, neither 'possession' nor 'capacity' works and I use 'condition'.

171. When the verb *prokheirizesthai* has *energeian* ('activity') as its direct object (13 times), I translate it 'mobilise', but when it occurs with *theôrian* (here and at 71,3) or *epistêmên* ('knowledge') (at 73,13), I prefer 'externalise' (for the externalisation of thought or knowledge, cf. 77,7-15), and when with *theôremata* ('theorems') (4 times), 'present'.

172. 'Thought' would be a reasonable rendering of *theôria* in this and the next chapter, but, for reasons explained in the note at 77,5, I have opted for the not altogether satisfactory 'reflection'. In fact, as 77,5-7 and 73,13 show, Philoponus could equally well have written 'externalising his knowledge of them', which would have been easier. Perhaps he does not because he has just used 'knowledge' and wants to avoid repetition.

173. Inserting *auta dekhesthai* after *pephukenai* at 46,19. (Rabe suggests inserting *dekhesthai*).

174. Changing *elegeto, ê ho* to *elegeto einai, ho* at 46,26-47,1, one of two possible corrections suggested by Rabe in the critical apparatus.

175. 46,17-22.

176. More literally 'not being active with respect to it'.

177. Changing *prôtên epitêdeiotêta* ('first receptivity') at 47,14, which is neither in accord with Philoponus' normal usage nor good sense – what would second receptivity be? – to *prôtên dunamin*. (Another possibility would be *phusikên epitêdeiotêta*, or 'natural receptivity; cf. 46,6, etc.).

178. 42,6-8.

179. Changing *ouketi* to *ouk esti* at 48,1, as suggested by Rabe in the critical apparatus.

180. Presumably commenting on *An. Post.* 98a35-b38. The commentary is not extant.

181. 42,4-8.

182. Changing *tôn* to *tou* at 49,18, as suggested by Rabe in the critical apparatus.

183. 42,13-14.

184. Inserting *to* before *axiôma* at 50,10, as suggested by Rabe in the critical apparatus.

185. 49,25-50,6.

186. 42,14-43,2.

187. 49,25-50,6 and 50,22-5.

188. Changing *tôn* to *tou* at 51,12, as suggested by Rabe in the critical apparatus.

189. 49,20-4.

190. Changing *autos* to *auto* at 51,24.

191. Changing *houtos* at 52,9 to *houtôs* rather than to *autos* as suggested by Rabe in the critical apparatus. (The reference of *houtôs* ('in that sense') will be to *kath' hexin* ('in capacity') two lines earlier).

192. More literally 'There being a builder by capacity'.

193. *Int.* 17a34-7.

194. A rather more literal translation of the phrase translated 'by purely verbal means' would be: 'by words and phrases'.

195. Rabe's suspicion of *ton* at 53,7 seems misplaced. When the first creator is only potentially a creator, the second creator is only potentially the cause of its becoming an actual creator and so it makes perfectly good sense to attribute the potentiality of the latter to that of the former.

196. In other words, even if one were to argue that it is the mental blueprint of a ship in the mind of the shipwright rather than the shipwright himself that brings the ship to actuality, that is no more an actual ship than is the shipwright.

197. Literally 'becomes such'.

198. 53,3-10.

199. Changing *tou dunamei auto* to *tou to dunamei autou* at 54,19. (cf. 53,21-2; 54,15-16; 55,14).

200. *Alc. 1* 106D.

201. The clause *dêlon oun ... agagôn* (54,24-55,1) seems to be defective in some way. For purposes of translation I have supplied after *autês* (54,25) the words *autos heautôi aitios ên tês ek tou proterou dunamei eis to energeiâi metabolês* ('was himself the cause of his transition from first potentiality to actuality') from 54,20-2, but it is not difficult to think of other possible supplements.

202. In *Metaph.* Aristotle states on a number of occasions that the universal is not a substance or a 'this' and in *An. Post.* that we gain a knowledge of it only by examining particulars, but he does not use the same language of it as Philoponus does here.

203. Two points need to be made about the translation of *akinêtos*. First, it shows the same kind of ambiguity as *genêtos* and could equally well be translated 'unmovable' or 'unmoved'. Because Proclus' *akinêton aition* is a direct descendant of Aristotle's unmoved mover, it is tempting to opt for 'unmoved', but 'unmoving' seems to work best as the argument unfolds and God and the universe are introduced. Second, *kinêsis*, which I normally translate 'movement', and related words such as *akinêtos* cover a wide range of 'movements', many of which are more naturally described as 'changes' in English and a good case could be made for translating *akinêtos* 'unchanging' and both *kinêsis* and the associated verb *kineisthai*, which occur later in the argument, 'change', but because Proclus goes on to say that being *akinêtos* implies being *ametablêtos*, and *ametablêtos* itself needs to be rendered by 'unchanging' or a synonym, I have opted for 'unmoving', etc. (For the range of 'movements' or 'changes' covered by *kinêsis*, see 256,19-22, where Philoponus in fact treats *kinêsis* and *metabolê*, or 'change', as synonyms).

204. That is, it will not come into being or pass out of existence. (Here in particular 'unchanging' might be thought a better rendering of *akinêtos* than 'unmoving').

205. Here, in the argument, and where possible in the refutation (for an exception, see the note at 62,7), I translate *poiein* 'produce' (and *poiêtês* 'producer', etc.). Another possible rendering, and one which would at times work better, would be 'create', but I reserve that for *dêmiourgein*, which also occurs with some frequency in the refutation.

206. Changing *aïdiou* to *aïdion* at 56,10.

207. The argument of this paragraph is close to that of Proclus *ET* 76,5-12.

208. I use 'perfect' / 'imperfect' to translate *teleios* / *atelês* when they describe God or human agents, 'complete' / 'incomplete' when they describe activities or movements.

209. On the question of the identity of this 'someone', see the Introduction.

210. The 'truth' referred to here is Christian doctrine, as is clear from 75,7-9 below. Other instances of *alêtheia* of which this may be so occur at 59,24; 61,8; 98,20; 117,16; 120,20; 127,3; 312,10.

211. Omitting *ou* at 57,15, as suggested by Rabe in the critical apparatus.

212. In the present refutation, and often in other works, Philoponus seems to use the comparative form of this adjective where one would expect the positive. I have not attempted to indicate this in the translation.

213. Changing *oude* at 57,21 to *ou dia* (cf. 75,21-2).

214. Changing *atelês ôn* to *atelê* at 59.4. (Rabe suggests *atelê auton*).

215. In *Phys.* 3.2 (especially 201b31-2).

216. At 60,21-2.

217. cf. 55,25-56,6 (especially the last three lines) – but the wording is much closer to that of Philoponus' own paraphrase of Proclus' argument at 59,26-7.

218. The antitheses *poiêsis* / *pêsis* and *poiein* / *paskhein* and the phrase *ouk auto poiei eis heteron* ('not itself acting on someone else') suggest that *poiêsis* and *poiein* here need to be translated 'acting' and 'act'. The charitable view is that the argument is shorthand for something like: (1) the change from first potentiality to the possession of a capacity is not a case of acting (*poiêsis*) but of being acted upon; (2) 'production' (*poiêsis*) is a case of acting (*poiêsis*); (3) therefore 'production' (*poiêsis*) cannot be an instance of a change from first potentiality to the possession of a capacity. However, it is difficult to avoid the suspicion that the argument actually depends on the ambiguity of *poiêsis* and *poiein* (the way having been prepared at 61,12-14) and that they are meant to retain (or at least include) the meanings 'producing' and 'produce'. (Two further points. (1) One might be tempted to ask oneself at this point whether Philoponus has been reading Proclus' *poiein* as 'act' rather than 'produce' all along, but *poiein* at 61,2 and *poiêsis* at 60,18 ff. show that this is not so. (2) How should one read *poiein* and related words in the rest of the section? I suspect that the answer is that the equivocation, once introduced, runs at least to the end of the section, but in the translation I have opted for 'produce' throughout).

219. 56,6-7.

220. 56,6-7.

221. Over the next four sections.

222. There is initially some temptation to translate *diastasis* 'extension', but 65,23-4 shows that this would be incorrect.

223. The passages shedding most light on this distinction between perfect and imperfect activity, or, as he sometimes puts it, between movement and activity or actuality, are *Phys.* 201b27-202a3; *Metaph.* 1048b18-36; 1065a14-1066a26; *DA* 431a6-7; *EN* 1174a13-b9.

224. *Phys.* 201a10-11.

225. He probably has in mind *DA* 431a4-7.

226. Rabe suggests supplying *tis* ('anyone') as the subject of the verb, but the reference seems to be to Proclus and his claim that an act of creation would involve God in movement.

227. *Phys.* 202a13-b29; cf. *DA* 417b2-9.

228. Punctuating with a comma rather than a full stop after *akhronos* at 66,18.

229. sc. minds.

230. Punctuating with a full stop rather than a comma after *kinêsis* at 66,25.

231. Or perhaps 'to quote *verbatim* some passages in which Aristotle testifies ...'.

232. 201a15-18.

233. 201b7-15.

234. 202a13-17.

235. 61,12-22.

236. 46,3-47,17.

237. 417a21-b2.

238. *theôrein* is difficult to translate. The activity described is, I think, that of attending to, or calling into consciousness, previously acquired knowledge, possibly, as the next clause would seem to suggest, in the course of applying it to a particular situation; or, to put it differently, active, as opposed to passive or potential, knowing. None of the usual renderings of *theôrein* really gets this across, and I have settled on 'reflect' (the choice of Smith in the *Revised Oxford Translation*), largely because 'reflection' works tolerably well as a rendering of *theôria* in what follows. (Although Philoponus' interpretation of the present passage differs in some respects from the one outlined above (see the note at 71,13), what he says at 77,5-10 shows that he too understands *theôrein* here as descriptive of a mental act which is identical with, or at least embodies, knowing).

239. After 'knowers' Ross (*Aristotle, De Anima* (Oxford, 1961)) adds *ontes energeiâi ginontai epistêmones* ('who become actual knowers') and Smith (op. cit.) follows his text.

240. Philoponus' paraphrase at 71,21 ff. shows that this is how he understood this phrase. Ross (op. cit. 233-4) and other modern scholars (e.g. Smith in the *Revised Oxford Translation*) take a different view of it. (I use 'condition' to render *hexis* here because ignorance can hardly be described as a capacity).

241. Or perhaps 'on the passage'.

242. 4.6.

243. Changing *tois* to *hois* and repositioning *pôs* to follow *adioristôs* at 70,11. The text printed by Rabe would read: 'The circumstances of the divisions of potentiality and of actuality are pretty clear from what has already been said by us in the last section; for since we have set out quite generally and without distinctions in earlier chapters [or perhaps "passages", or "arguments"] how Aristotle has used "potentiality" and "actuality" (he indicates this by writing "for up until now we have been talking of them quite generally"), he carries out a division of them in the passage we have quoted'. This obviously cannot stand. In particular, the shift from 'we' to 'he' in the second part of the sentence is clearly unacceptable. (A case could be made for more radical surgery. One might, for example, change *tôn êdê hêmin proeirêmenôn en tôi pro toutou kephalaiôi* at 70,9-10 to, say, *tôn legomenôn* and omit the words *parethêkamen pôs* – or perhaps just *parethêkamen* – at 71,11 and translate: 'The circumstances of the divisions of potentiality and of actuality are pretty clear from what is said; for since Aristotle has used "potentiality" and "actuality" in a rather general and loose manner in earlier passages (he indicates this by writing "for up until now we have been talking of them quite generally"), he carries out a division of them in the passage we have quoted'. Although this is remote from the transmitted text, someone who failed to see that Philoponus is moving straight into the promised elucidation of Aristotle's words could have revised the passage under the influence of 69,7-11).

244. 69,13-14.

245. For 'externalise' and 'reflection', see the notes at 46,14 and 15.

246. It should be noted that the above exegesis of Aristotle's words, together with what he says elsewhere (see especially 3.2 and 77,5 ff.), shows that for Philoponus the three phases in Aristotle's illustration of the potentiality / actuality contrast are (1) the potential for acquiring knowledge; (2) the possession of (latent) knowledge; (3) the 'mobilisation' of knowledge (which appears to mean its 'externalisation' or presentation or use in the world), whereas on my understanding of them (for which see the note at 69,20) they would be (1) the potential for acquiring knowledge; (2) the possession of (latent) knowledge; (3) conscious, or active, knowing. (Under the first of these schemes *theôria* is a virtual synonym for 'knowledge' (see 77,5-7), and (1) is a potentiality, (3) an actuality, and (2) an actuality relative to (1) but a potentiality relative to (3); under the second, *theôria* is the 'conscious, or active knowing' of (3), and (1) and (2) are different kinds of potentiality, and (3) an actuality).

247. Adding *ho* after *hoti* at 71,15.

248. 69,22-5.

249. *eis epistêmên* is rather odd; perhaps Philoponus actually wrote either *kata epistêmên* or *eis epistêmôna* ('into a knower').

250. The Greek here could be rendered 'something which', but later at 72,9 the Greek, rather oddly perhaps, has a masculine rather than a neuter form.

251. Perhaps he has 5.1-2 in mind, and in particular 224a30-b8 and 225b23-4 for alteration as change of form and 225a12-20 and b3-5 (which, incidentally, seems to contradict Philoponus' claim) for the change from privation to form.

252. Changing *kai thermou de homoiôs kai psukhrou* at 72,11-12 to *kai ho thermainomenos de homoiôs ek psukhrou*, which is, apart from the addition of *ho*, a correction suggested by Rabe in the critical apparatus.

253. i.e. the coming into existence of a new substance rather than a qualitative change.

254. *DA* 417b5-9.

255. See the note at 69,20 on the difficulties of translating *theôrein*. Smith in the *Revised Oxford Translation* has 'For what possesses knowledge becomes an actual knower', which, although not literal, doubtless catches the sense.

256. 72,4-8.

257. 61,5 ff. The whole section should be compared with this one.

258. Changing *orthotêtos* at 74,26 to *deinotêtos*. (*orthotês* should mean 'correctness' or 'rectitude' and could conceivably stand if one assumes that Philoponus is being ironic. However, Philoponus wrote *deinotêtos* rather than *orthotêtos* at 61,9 in the passage to which he has just directed the reader and it seems likely that that is what he wrote here).

259. *James* 1,17.

260. Until now I have been translating *epistêmê* 'knowledge', but it can also be rendered 'skill' or 'science' and on balance 'skill' seems to work best in this section.

261. *iskhein*. Perhaps 'maintain' would be more literal.

262. Or 'contemplating and thinking'. *theôria* ('contemplation') clearly owes its presence here to Aristotle's use of *theôrein* in the passage from *DA* that was cited in section 7 and *noêsis* ('thought') is in effect an interpretative gloss on it. In what follows, Philoponus uses the two words interchangeably, and often 'thought' or 'thinking' would be the most satisfactory rendering of *theôria*, but I have thought it best to maintain a distinction between the two words in the translation.

263. Or possibly 'ideas' (cf. Lampe s.v. A).

264. More literally 'we are then said to be active with respect to the capacities'.

265. i.e. not clothed in words.

266. sc. in someone other than the thinker.

267. Although I have used personal forms in the translation, the entities that reflect and communicate are, from 'reflecting' to the end of the sentence, impersonal (i.e. in the neuter gender) in the Greek.

268. 2.5.

269. 13,26-23,22.

270. Philoponus returns to the question of the nature of God's will in 16.1-4.

271. Philoponus appears to use the verbs *boulesthai* and *thelein* interchangeably in this section and I shall not attempt any systematic distinction between them in the translation.

272. Changing *eis to autou tou Sôkratous sôma* to *eis to auto tôi tou Sôkratous sômati* at 79,26. (cf. LSJ *autos* III.1 and the synonymous phrase *einai meta tôn sômatôn* at 80,23; Rabe writes '*einai eis* suspecta' in the critical apparatus).

273. i.e. the second negative cancels, or reverses, the first.

274. 41C-D.

275. Changing *ho autos nous hôste tas* to *ho autos anthrôpos nosôn te tas* at 81,20-1, as suggested by Rabe in the critical apparatus, and further adding *ê* after *hugiainôn* at 81,21.

276. As patients rather than agents.

277. Changing *dêmiourgêsei* to *dêmiourgêsai* at 82,6, as suggested by Rabe in the critical apparatus.

278. 56,11-12.

279. 56,18-20.

280. Removing the existing diacritics from *tauta* at 82,15 and accenting it with a circumflex on the first syllable (cf. 96,2).

281. Rabe may be right in thinking that something has gone wrong with the Greek text at this point, but the sense is clear enough.

282. He presumably has in mind Anaxagoras' doctrine that there is a portion of everything in everything (cf. Diels-Kranz frs. B6 and B11).

283. In the case of Plato the reference must be to *Phd.* 72A-D; Aristotle criticises the relevant doctrine of Anaxagoras in a number of places, in most detail in *Phys.* 1.4.

284. More literally 'even though no perception grasps it'.

285. Whether by being low, mean and trivial, or by being sketchy and undeveloped. In the former case he would have people like the Epicureans in mind, in the latter the uneducated.

286. i.e. 'will be heir to all the ills of the body' – and so not be perfect and free of need.

287. *DA* 417b5-16.

288. The original has 'that'.

289. *Epist.* 2, 312E.

290. *Laws* 715E.

291. *Il.* 2. 204; *Metaph.* 1076a4.

292. In the previous sentence.

293. *Phys.* 256a5-8.

294. 256a8-13.

295. The first few lines of the quotation differ markedly from the text of Aristotle and seem to be somewhat garbled. I have, however, resisted the temptation to fall back on the text of Aristotle and, apart from changing *ekeino* to *ekeina* at 89,14, have done my best with the transmitted text. (Aristotle has: 'Now we say that both the first and the last of the movers imparts motion, but more so the first, for it moves the last but not it the first, and without the first ...'.)

296. *Phys.* 257a10-12.

297. 257b15-20.

298. Changing *aitiôtaton* to *aitiôteron* at 90,5, as suggested by Rabe in the critical apparatus.

299. Retaining *estin*, the manuscript reading, at 90,10 and changing *tôi aitiôi tês kinêseôs estin* to *tôi aitiôteran tês kinêseôs einai* at 90,11-12.

300. 5,15-18 (Boese) (= 5,21-4 (Isaac)). Also quoted at 38,3-6 and 570,1-4.

301. 5,22-5 (Boese) (= 5,30-4 (Isaac)). Also quoted at 38,10-15 and 570,8-13.

302. 5,32-5 (Boese) (= 5,43-8 (Isaac)). Also quoted at 6,17-21; 38,16-20 and 570,14-18.

303. Adding *aei* before *estin* at 92,5, as suggested by Rabe in the critical apparatus.

304. 42,12-43,15.

305. Omitting *dunamei kai* after *deuteros* at 92,18.

306. Changing *ê* at 92,20 to *ei*.

307. Changing *ê* at 92,22 to *eiê*.

308. 41A.

309. The meaning (and, consequently, the correct translation) of the words translated 'Gods, offspring of gods, works of which I am the creator ...' is far from clear. The rendering adopted here is unlikely to represent Plato's intention but has a better chance of reflecting Philoponus' understanding of it. (Anyone interested in the issues involved should consult F.M. Cornford, *Plato's Cosmology, the* Timaeus *of Plato* (London, 1937), 367-70).

310. 41B-C.

311. At 42,4-6. One might have expected *epikheirêmati* rather than *kephalaiôi*, but see the note at 1,14.

312. 43,16-17.

313. 82,13.

314. Inserting *dedêmiourgêtai kai aei* after *aei* at 95,7, as suggested by Rabe in the critical apparatus.

315. i.e. that he will be trapped by his own arguments. The literal meaning of the Greek is 'that he will be caught by his own feathers' and the reference is to eagles shot with arrows feathered with their own plumes. The saying, which appears in some fifteen Greek authors, including Philoponus' critic Simplicius, goes back to the dramatist Aeschylus (Fr. 139) and the Shakespearean phrase is, as LSJ suggests, a good equivalent.

316. Plato *Tim.* 34A.

317. Although 96,20-1 shows that this is how Philoponus understands this clause, I think that Proclus intended something like 'for whatever is there [sc. in time], is in a particular [part of] time'.

318. Changing *einai* in line 5 to *esti* and (1) closing the brackets after it rather than after *thraxeien* in line 4 (2) changing the full stop after *hêmin* in line 6 to a comma (3) removing the brackets around *hoson* ... *meros* (lines 7-9), and punctuating instead with a semicolon after *estai* in line 7 and a full stop after *meros* in line 9.

319. The words *hoson* ... *ou pote* ('for as much as ... not at some [particular] time') are quoted again at 167,17-20, but there the words *hoson* ... *meros* ('for as much as ... bit by bit') are placed after *estin men pote* ('not at some [particular] time').

320. Changing *dia ti* at 96,19 to *dioti* (cf. 99,2), and punctuating with a comma rather than a question mark at 96,21.

321. Philoponus probably has passages such as 82,15-25 and 95,5-12 in mind, but much in Proclus' second, third and fourth proofs might be thought to imply the same principle.

322. 94,23-5.

323. Changing *pote* at 98,11 to *ton te*. (In the critical apparatus Rabe raises the possibility of inserting *ton te* but not at the expense of *pote*.)

324. At 96,2-18.

325. 96,6-9.

326. For the translation of this last clause, see the note at 96.8.

327. 96,9-10.

328. 96,11-13.

329. 96,13-14.

330. 98,23-5.

331. 96,14-17.

332. Changing *genesthai* to *ginesthai* at 99,19, as suggested by Rabe in the critical apparatus.

333. Changing *anankê* to *anankazei* at 100,18. (Rabe suggests a more complex correction in the critical apparatus).

334. Changing *tou* ... *parêgmenou* at 101,2-3 to *tôn* ... *parêgmenôn* (cf. 85,5; 87,1; 88,10).

335. Literally 'more imperfect than himself'.

336. Changing *hôsper* to *eiper* at 101,6.

337. At 39,5-18. (The passage is quoted for a third time at 571,20-572,8).

338. On this title, see the note at 39,3.

339. *Enn.* 6.7.1,45-57.

340. Although Proclus frequently uses *epi toutois* with an ordinal number in this sense in lists of points in other works, this is the only time he uses it at the beginning of an argument in *Aet.*

341. In what follows I use 'exist' and 'be' interchangeably to translate the verb *einai*.

342. Or 'at the same time with', but it seems best to avoid an obvious temporal connotation.

343. Thomas Taylor and Lang and Macro assume that *ouranos* ('the heaven') is the unexpressed or lost subject of the second *ên* in line 5, but I believe, like Philoponus (see, for example, section one of the refutation) and Baltes (see his translation at *Die Weltentstehung*, 2, 139) that the part of the argument that runs from line 4 to line 18 depends entirely on the supposed implications of talking of 'when' time did not, or will not, exist. If this is right, the present sentence would seem to read 'But there was no time when it [sc. time] was not and will be [none] when it will not be', which is unexpected because later in the argument (on my reading of it) Proclus studiously avoids seeming to pre-empt the argument by talking directly of 'a time (*khronos*) before time (*khronos*)' by using such unnatural locutions as *ên hote ouk ên* ('there was a "when" there was not ...'; see lines 5, 10 and 13; I use 'a when' in translating these expressions so as to avoid the more natural rendering 'a time when', which would risk confusion). I suspect (and have translated accordingly) that the solution to the difficulty, if a difficulty it is, is that *khronos de ouk ên hote ouk ên* is an inversion of *ou de ên hote khronos ouk ên* brought about by a desire to avoid the juxtaposition *ou de*, which Proclus, like many other writers (cf. J.D. Denniston, *The Greek Particles* (Oxford, 1934), 187), normally shuns. (A TLG search of Proclus' oeuvre shows only two instances of *ou de*, both of which could be written *oude*). The subject of the second *ên* in line 5 would then be expressed and there would be no possibility of understanding *ouranos*. (If I am wrong about there being an inversion, I would supply *khronos* as the subject of *ên* and translate 'But there was no time when [time] did not exist and there will be none when it does not exist' but harbour a suspicion that something had gone wrong with the text).

344. *pote* ('some time') is an indefinite adverb of time closely related in form and meaning to the relative adverb of time *hote* ('when'); to talk of 'when' a thing exists is to talk of it existing 'at some time'. Unfortunately, I have not been able to find a satisfactory English representation of *pote* which does not include the word 'time'. In what follows (both in the argument and in the refutation) the reader needs to be aware that 'time' on its own normally translates *khronos* and that phrases such as 'some time', 'at some time', 'at one time', 'at another time' normally translate *pote*.

345. sc. 'is a time expression'; or perhaps 'is [a part] of time'.

346. Adding *aei* after *estin* at 103,25, as suggested by both Thomas Taylor and Rabe. (According to Baltes the need for the addition is confirmed by Is<u>h</u>âq ibn <u>H</u>unayn's Arabic translation).

347. *Tim.* 38C.

348. *perixeei*. The word, which is uncommon, actually means 'polish all around', and I have not found other instances of its being used figuratively.

349. More literally 'weakness'.

350. More literally 'of whatever it may be predicated'.

351. Changing *tina huparxin* to *tinos huparxeôs* at 104,16 (cf. 116,1-22, especially 14-15 and 20).

352. Many of the arguments Philoponus uses in the first four sections of the refutation are also found at *in Phys.* 456,17-458,32, which is translated in M.J. Edwards (tr.), *Philoponus: On Aristotle's* Physics 3 (London & Ithaca, NY, 1994), 119-21.

353. Philoponus uses the phrase *ên pote hote ouk ên* ('there was some time when ... did not exist') on a number of occasions both here and in *in Phys.* Since neither Proclus nor Aristotle actually uses it (although Proclus comes close) and he can hardly have been unaware that it had been used in controversy by the Arian heretics and was one of a number of phrases anathematised by the church at the Council of Nicea in 325 (cf. Edwards, op. cit., 171, n. 206), I am tempted to believe that Christian readers were meant to read its attribution to Proclus here as a subtle anathematisation of his position. (The more direct relevance of the arguments in which it occurs to

Proclus' argument than to anything in the *Physics* suggests that they were developed for *Aet.* and reused in *in Phys.*, which would be consistent with Verrycken's argument (see 'The Development of Philoponus' Thought and Its Chronology', 244-54) that the latter was revised after 529).

354. Changing the first *ên* to *êi* at 104,27, as suggested by Rabe in the critical apparatus.

355. At *Il.* 20. 250.

356. At 103,15-16.

357. *sullabai* are normally syllables and the image here could conceivably be one of young children breaking words up into their syllables to create nonsense words, but a *meirakion* is normally an adolescent or a young man, and at *in Phys.* 508,10-11 Philoponus writes that the elements of a syllogism are its premisses and those of a premiss its *sullabai*, by which he presumably means its component terms rather than their syllables.

358. For the history of the use of time words in non-temporal senses in ancient philosophy see R. Sorabji, *Time, Creation and the Continuum*, 112-16.

359. Changing *pausetai* to *pauetai* at 107,21, as suggested by Rabe in the critical apparatus. (Both verbs should be of the same tense and 107,25-6 favours the present).

360. One would expect something like *to pote palin mê noein* rather than *tote palin noein* at 108,12, but Philoponus himself may have lost sight of the construction and I have done my best with the text printed by Rabe.

361. *Tim.* 29E.

362. Literally 'yet more'.

363. cf. Aristotle *Phys.* 219b1-2; Plotinus *Enn.* 3.7.9.

364. The Greek actually has 'since'.

365. Or perhaps, 'interpreting "will be" as a time'; or even 'changing "will be" into a time'. (The same phrase occurs at 112,1 and 113,1).

366. Adding *gar* after *ean* at 111,27, as suggested by Rabe in the critical apparatus.

367. Deleting *ouk* at 112,3.

368. Deleting *ouk* at 112,5.

369. Adding *ti* after *atopon* at 112,17, as suggested by Rabe in the critical apparatus.

370. Omitting *kai* after *tmêmata* at 113,18, as suggested by Rabe in the critical apparatus.

371. Changing *temnomenos* to *temnomenon* at 115,2, as suggested by Rabe in the critical apparatus.

372. *Tim.* 38B-C.

373. I have restored the words *kai kata to paradeigma tês aiôniou phuseôs* from the text of Plato after *ginêtai* at 115,17. Without them, the next phrase would have to be construed 'so that [time] may be as similar as possible to [the heaven]'. This can hardly have been what Philoponus intended, and the missing words are in fact present when he quotes the passage again at 141,1-7 and at 554,10-16, so it seems likely that the phrase has dropped out during the transmission of the text.

In the Greek neither the subject of *êi* at 115,18 nor the reference of *ho d'* at 115,19 is immediately clear. Among modern translators, Jowett opted for 'time' and 'the heaven' respectively, Lee (if I read him correctly) for 'time' in both places, and Cornford (see *Plato's Cosmology*, 99, n. 1) for 'the heaven' in both places. Although certain aspects of what Philoponus goes on to say in this chapter might be thought to imply that he understood 'time' in both places, when he quotes the present lines again at 554,10-22 he goes on to paraphrase them in a way that makes it clear that he actually believed that Plato is talking about the *kosmos* at both points, and I have translated accordingly. (My own view is that Cornford is right when he claims that comparison with *Tim.* 37C8 and 39E1 suggests that the subject of *êi* should be *ouranos*, but that in the present state of the text one inevitably reads it as *khronos*. Accordingly, I would argue that *ho ouranos* should be restored after *êi* in the text of the *Timaeus*).

374. Adding *ton aiôna* after *zôês* at 115,22. Accepting Rabe's text, one would have to translate either: 'And so, if Plato says that the whole of infinity is the pattern for the world and the measure of the life of everlasting things ...', or: 'And so, if Plato says that the pattern for the world has being for all eternity and is the measure of the life of everlasting things ...'. The first translation would not only make infinity the pattern for the world and not just for time, but would radically change the status of the phrase *panta ton aiôna*, which is clearly adverbial in the *Timaeus* passage; the second would make the pattern for the world, rather than infinity, 'the measure of the life of everlasting things'. (A comment of John Bowin, one of the editors of this volume, pointed me in the right direction here).

375. 114,20-116,1 are translated and discussed by R. Sorabji in *Time, Creation and the Continuum* (London & Ithaca, NY, 1983), 117-19.

376. Or perhaps 'there was a time'.

377. *Tim.* 29E.

378. Changing *khronikon* to *khronikôs* at 116,14, one of three possible corrections suggested by Rabe in the critical apparatus.

379. 106,6-14.

380. The construction is such that 'the truth' is coordinate with 'Plato' with the result that it is almost personified and there may be a reference to Christian scripture, or more strictly speaking to the 'truth' embodied in it. If this is so, such a reference would be a little surprising in what seems to be a summary of what has gone before, since there has been no direct reference to the Christian position earlier in the argument, and the words *kai têi alêtheiâi* could be an interpolation.

381. Adding *legonta* after *Platôna* at 117,19, as suggested by Rabe in the critical apparatus.

382. See the note at 9,22.

383. *Tim.* 38B (time); 28B (the heaven).

384. 251b14-19.

385. Translating Aristotle's *exô henos* rather than the *ex aiônos* of M, which Rabe retains. (Although M also has *ex aiônos* at 218,22, where the passage is quoted again, the substitution is more likely to be due to a copyist than to Philoponus).

386. Arist. *Phys.* 251b14-19 (with minor differences).

387. In sections 7-29.

388. Changing *dei* to *dein* at 118,17, as suggested by Rabe in the critical apparatus.

389. Adding *logon* after *toiouton* at 118,19, one of two possible corrections suggested by Rabe in the critical apparatus.

Select Bibliography

The following is a list of the principal works cited in the notes.

Anawati, G., 'Un fragment perdu du *de Aeternitate mundi* de Proclus', in *Mélanges de philosophie grecque offerts a Mgr. Diès* (Paris, 1956), 21-5.

Armstrong, A.H. (tr.), *Plotinus* (Cambridge, Mass., 1966-88).

Badawi, A., *Neoplatonici apud Arabes* (Cairo, 1955).

Baltes, M., *Die Weltentstehung des platonischen* Timaios *nach den antiken Interpreten* (Leiden, 1976).

Baltzly, D., Review of H.S. Lang and A.D. Macro, (trs), *On the Eternity of the World* (De Aeternitate Mundi), *Proclus, Bryn Mawr Classical Review*, 2002.10.19.

Bastid, P., *Proclus et le crépuscule de la pensée grecque* (Paris, 1969).

Boese, H. (ed.), *Procli Diadochi Tria opuscula de providentia, libertate, malo* (Berlin, 1960).

Cornford, F.M. (tr.), *Plato's Cosmology, the* Timaeus *of Plato* (London, 1937).

D'Ancona Costa, C., 'Plotinus and later Platonic philosophers on the causality of the first principle', in L.P. Gerson (ed.), *The Cambridge Companion to Plotinus* (Cambridge, 1996).

De Groot, J., *Aristotle and Philoponus on Light* (New York & London, 1991).

Denniston, J.D., *The Greek Particles* (Oxford, 1934).

Des Places, E. (ed.), *Atticus: fragments* (Paris, 1977).

Diels, H. and W. Kranz (eds), *Die Fragmente der Vorsokratiker*, 6th ed. (Berlin, 1951-2).

Dillon, J., *The Middle Platonists: a Study of Platonism, 80 BC to AD 220* (London, 1977).

Edwards, M.J. (tr.), *Philoponus, On Aristotle's* Physics 3 (Ithaca, NY, 1994).

Guthrie, W.K.C., *A History of Greek Philosophy*, vol. 1 (Cambridge, 1967).

Hankinson, R.J., *Cause and Explanation in Ancient Greek Thought* (Oxford, 1998).

Hoffmann, P., 'Simplicius' polemics: some aspects of Simplicius' polemical writings against John Philoponus: from invective to a reaffirmation of the transcendency of the heavens', in R. Sorabji (ed.), *Philoponus and the Rejection of Aristotelian Science* (London & Ithaca, NY, 1987), 57-83.

Hussey, E. (tr.), *Aristotle's* Physics *Books II and IV* (Oxford, 1983).

Isaac, D. (ed.), *Dix problèmes concernant la providence* (*Proclus, Trois études sur la providence*, vol. 1) (Paris, 1977).

Jowett, B. (tr.), 'Timaeus', in E. Hamilton and H. Cairns (eds.), *The Collected Dialogues of Plato* (Princeton, NJ, 1961).

Judson, L., 'God or nature? Philoponus on generability and perishability', in R. Sorabji (ed.), *Philoponus and the Rejection of Aristotelian Science* (London & Ithaca, NY, 1987), 179-96.

LSJ, see Liddell, H.G.
Lampe, G.W.H., *A Patristic Greek Lexicon* (Oxford, 1961).
Lang, H.S. and A.D. Macro (trs), *On the Eternity of the World* (De Aeternitate Mundi), *Proclus* (Berkeley, LA & London, 2001).
Lee, H.D.P. (tr.), *Plato, Timaeus* (Harmondsworth, 1965).
Liddell, H.G. and R. Scott (comps.), *A Greek-English Lexicon*, rev. H. Jones; with a New Supplement (Oxford, 1996).
Ostwald, M. (tr.), *Nicomachean Ethics* (Englewood Cliffs, NY, 1962).
Owens, J., *The Doctrine of Being in the Aristotelian 'Metaphysics': a Study in the Greek Background of Mediaeval Thought*, 2nd ed. rev. (Toronto, 1963).
The Oxford Classical Dictionary, 3rd ed.; ed. by S. Hornblower and A. Spawforth (Oxford, 1996).
Pinès, S., 'An Arabic summary of a lost work of John Philoponus', *Israel Oriental Studies* 2 (1972), 320-52; repr. in S. Pinès, *Studies in Arabic Versions of Greek Texts and in Medieval Science*, vol. 2 (Leiden & Jerusalem, 1986), 294-326.
Rabe, H. (ed.), *Ioannes Philoponus de Aeternitate Mundi contra Proclum* (Leipzig, 1899) (repr. Hildesheim, etc. 1984).
Rackham, H. (tr.), *The Nicomachean Ethics* (London & Cambridge, Mass., 1934).
Rosenthal, F., 'From Arabic books and manuscripts VII: some Graeco-Arabica in Istanbul', *Journal of the American Oriental Society* 81 (1961), 7-12.
Ross, W.D. (ed.), *Aristotle, De Anima* (Oxford, 1961).
Ross, W.D. (tr.), *Metaphysics*, in J. Barnes (ed.), *The Complete Works of Aristotle: the Revised Oxford Translation* (Oxford, 1984).
Ross, W.D. (tr.), *Nicomachean Ethics*, in J. Barnes (ed.), *The Complete Works of Aristotle: the Revised Oxford Translation* (Oxford, 1984).
Saffrey, H.-D., 'Allusions antichrétiennes chez Proclus: Le Diadoque platonicien', *Revue des sciences philosophiques et théologiques* 59 (1975), 553-63.
Saffrey, H.-D. and L.G. Westerink (eds.), *Proclus, Théologie platonicienne* (Paris, 1968-94).
Sambursky, S., *The Physical World of Late Antiquity* (London, 1962).
Scholten, C., *Antike Naturphilosophie und christliche Kosmologie in der Schrift 'De Opificio Mundi' des Johannes Philoponus* (Berlin & New York, 1996).
Schroeder, F.M., 'Plotinus and language', in L.P. Gerson (ed.), *The Cambridge Companion to Plotinus* (Cambridge, 1996).
Siorvanes, L., *Proclus: Neo-Platonic Philosophy and Science* (New Haven & London, 1996).
Smith, J.A. (tr.), *On the Soul*, in J. Barnes (ed.), *The Complete Works of Aristotle: the Revised Oxford Translation* (Oxford, 1984).
Sorabji, R. (ed.), *Aristotle Transformed: the Ancient Commentators and Their Influence* (Ithaca, NY, 1990).
Sorabji, R., *Matter, Space and Motion: Theories in Antiquity and Their Sequel* (London & Ithaca, NY, 1988).
Sorabji, R. (ed.), *Philoponus and the Rejection of Aristotelian Science* (London & Ithaca, NY, 1987).
Sorabji, R., *Time, Creation and the Continuum: Theories in Antiquity and the Early Middle Ages* (London & Ithaca, NY, 1983).
TLG, see *Thesaurus linguae graecae.*

Tarán, L., 'Amicus Plato sed magis amica veritas, from Plato and Aristotle to Cervantes', *Antike und Abendland* 30 (1984), 93-124.

Taylor, T. (tr.), *The Fragments That Remain of the Lost Writings of Proclus, Surnamed the Platonic Successor* (London, 1825) (repr. San Diego 1988).

Thesaurus linguae graecae [CD ROM].

TrGF, see *Tragicorum Graecorum fragmenta.*

Tragicorum Graecorum fragmenta, vol. 3 (*Aeschylus*), ed. S. Radt (Göttingen, 1985).

Verrycken, K., 'The development of Philoponus' thought and its chronology', in R. Sorabji (ed.), *Aristotle Transformed: the Ancient Commentators and Their Influence* (London & Ithaca, NY, 1990).

Verrycken, K., 'Philoponus' interpretation of Plato's cosmogony', *Documenti e Studi sulla Tradizione Filosofica Medievale* 8 (1997), 269-318.

Westerink, L.G. (ed.), *The Anonymous Prolegomena to Platonic Philosophy* (Amsterdam, 1962).

Wildberg, C. (tr.), *Philoponus: Against Aristotle on the Eternity of the World* (London & Ithaca, NY, 1987).

Wildberg, C., 'Prolegomena to the study of Philoponus' contra Aristotelem', in R. Sorabji (ed.), *Philoponus and the Rejection of Aristotelian Science* (London & Ithaca, NY, 1987).

Wildberg, C. (tr.), 'Simplicius: against Philoponus on the Eternity of the World', in *Place, Void, and Eternity* (London & Ithaca, NY, 1991).

Williams, C.J.F. (tr.), *Aristotle's* de Generatione et Corruptione (Oxford, 1982).

Wisnovsky, R., 'Yaẖyâ al-Naẖwî', in *Encyclopaedia of Islam*, vol. 11 (Leiden, 2002).

English-Greek Glossary

able to see: *horatikos*
absurd: *atopos*
absurdity: *atopia, to atopon*
accept: *homologein, sunkhôrein*
accident: *to sumbebêkos*
accomplish: *teleioun*
account (n.): *logos*
accuse of: *katêgorein*
act: *energein*
acted upon, be: *paskhein*
active, be: *energein*
activity: *energeia*
actual: *energeiâi, kat' energeian*
actualisation: *entelekheia*
actuality: *energeia, entelekheia*
actually: *energeiâi, entelekheiâi, kat' energeian*
ad infinitum: *eis apeiron, ep' apeiron*
add: *epagein*
admit: *sunkhôrein*
affection: *pêsis*
affirm: *apophainesthai, kataphaskein*
affirmation: *kataphasis*
agent: *to poioun*
agree: *homologein, sunkhôrein*
air: *aêr*
all at once: *athroos*
also exist: *sunuparkhein*
alter (trans.): *alloioun, metaballein*
alteration: *alloiôsis*
always: *aei*
ambiguity: *homônumia*
ambiguous: *homônumos*
ambiguously: *homônumôs*
animal: *zôion*
apprehend: *noein*
archetype: *arkhetupon*
argue: *kataskeuazein, sunagein*
argue unfairly: *kakourgein en tois logois*
argument: *apodeixis, logos, sullogismos*
ask: *zêtein*
assert: *apophainesthai*
association: *koinônia*
assume: *hupotithenai, lambanein*
at all: *holôs*
at rest, be: *êremein*
attack: *prosballein*
attribute (n.): *to sumbebêkos*
authority: *exousia*
axiom: *axiôma*

be: *huphistasthai*
become: *ginesthai*
beginning: *arkhê*
beginningless: *anarkhos*
being: *ousia, ousiôsis*
belief: *doxa*
believe: *pisteuein*
belong: *huparkhein*
better (a.): *kreittôn*
bind to: *sundein*
birth: *genesis*
blasphemous: *blasphêmos*
blasphemy: *blasphêmia*
body: *sôma*
born together with: *homogonos*
bright: *phôteinos*
bring: *paragein*
bring into existence: *huphistanai*
bring to maturity: *teleioun*
bringing into being: *ousiôsis*
build: *oikodomein*
buildable: *oikodomêtos*
builder: *oikodomos*
building (sc. the process): *oikodomêsis, oikodomia*
building: *oikia*

by a creative act: *dêmiourgikôs*

call: *onomazein*
can: *endekhesthai*
capable of: *epitêdeios*
capable of coming to be: *genêtos*
capable of producing: *poiêtikos*
capable of sight: *horatikos*
capable of speech: *logikos*
capacity: *hexis*
cast a shadow over: *episkiazein*
cause: *aitia, to aition*
cause of: *aitios*
cause to exist: *huphistanai*
caused, that which is: *to aitiaton*
cease: *pauesthai*
cease to exist: *phtheiresthai*
change (v.: intrans.): *metaballein*
change (n.): *metabasis, metabolê*
characterise: *kharaktêrizein*
characteristic: *idiotês*
cite: *paratithesthai*
civic: *politikos*
clear (a.): *enargês*
clearly: *enargôs*
co-everlasting: *sunaïdios*
coexist: *suneinai, sunuparkhein, sunuphistasthai*
coexistent with, be: *sumparateinesthai, sunuparkhein*
cognitive: *gnôstikos*
colour (n.): *khrôma*
come to be: *ginesthai*
coming into existence: *huparxis*
comment (n.): *exêgêsis*
commentary: *hupomnêmata*
commentator: *exêgêtês*
common: *koinos*
common feature: *koinotês*
compel: *anankazein*
complete (v.): *sumplêroun*
complete (n.): *teleios*
component of, be a: *sumplêroun*
concede: *homologein, sunkhôrein*
conceive of: *ennoein, noein*
concept: *noêsis*
conception: *ennoia*
conclude: *sullogizesthai*
conclusion, to draw a: *sumperainein*
concord: *sumpnoia*
condition: *hexis*
consequence: *to hepomenon*
consist: *sunkeisthai*
consistent (with): *akolouthos*
constitute: *sumplêroun*
constituting: *sumplêrôtikos*
constitutive: *sumplêrôtikos*
construct: *sunistanai*
contentious: *philoneikos*
contentiousness: *philoneikia*
continuity: *sunekheia*
continuous: *sunekhês*
continuous becoming: *aeigenesia*
contradiction: *antiphasis*
contrary (a.): *enantios*
contrast (v.): *antidiastellein*
contribute to: *suntelein*
copy (n.): *eikôn*
corporeal thing: *sôma*
count off: *aparithmein, exarithmein*
count out: *exarithmein*
counter: *to arithmoun*
create: *dêmiourgein*
create fallacious arguments: *paralogizesthai*
created thing: *dêmiourgêma*
creating: *poiêsis*
creation: *dêmiourgêma, dêmiourgia, to genêton, hupostasis*
creation of substance: *ousiôsis*
creative: *dêmiourgikos*
creative activity: *dêmiourgia*
creator: *ho dêmiourgêsas, dêmiourgos, poiêtês*
creature: *zôion, to genêton*
criticise: *apelenkhein*
criticism: *elenkhos*

darkness: *skotos*
death: *thanatos*
decay (v.): *phthinein*
decay (n.): *phthisis*
declaration: *apophansis*
declare: *apophainein*
deduce: *sullogizesthai*
define: *horizesthai*
definite: *hôrismenos*
definition: *horismos, horos, logos*
demolish: *anairein, anaskeuazein*
demon: *daimôn*
demonstrate: *apodeiknunai, deiknunai*
demonstration: *apodeixis*
deprived of, be: *stereisthai*

destroy together: *sunanairein*
determine: *aphorizein*
devoid of, be: *stereisthai*
dialectical: *dialektikos*
didactic: *didaskalikos*
difference: *diaphora*
different: *alloioteros, diaphoros*
differentiation: *diaphorotês*
disperse: *merizein*
dissolution: *lusis*
dissolve: *luein*
distinguish: *diairein, prosdiorizesthai*
divide: *merizein*
divide up along with: *sundiïstanai*
divine: *theios*
divisible: *meristos*
division: *diairesis, diastasis*
do away with: *aphaireisthai*
doctrine: *dogma, doxa*
duration: *diastêma*
dwelling: *oikos*

earth: *gê*
eclipse (n.): *ekleipsis*
effect (n.): *to aitiaton, pathos*
element: *stoikheion*
eliminate: *anairein*
eliminate along with (also): *sunanairein*
empty talk: *kenologia*
end (n.): *peras, telos*
endless: *ateleutêtos*
envy (n.): *phthonos*
equivalent, be: *isodunamein*
essence: *ousia*
essence, as part of the: *ousiôdôs*
essence, of the: *ousiôdês*
essential: *ousiôdês*
establish: *kataskeuazein*
eternal: *aiônios*
eternally: *aiôniôs*
eternity: *aiôn*
ethical: *êthikos*
everlasting: *aïdios*
everlastingly: *aïdiôs*
everlastingness: *aïdiotês*
evil: *ponêros*
examine: *skopein*
example: *hupodeigma*
exercise providence: *pronoein*
exist: *huparkhein*
exist together (intrans.): *sunuphistasthai*
existence: *hupostasis, huparxis*
existent: *huparxis*
explain: *exêgeisthai*
explanation: *aitia*
explanatory: *hupomnêmatikos*
explicitly: *epi lexeôs*
exposition: *didaskalia*
expression: *dêlôsis, sêmasia*
extension: *ektasis, paratasis*
externalise: *prokheirizesthai*

facts, the: *enargeia, pragmata*
fall upon: *prosballein*
fallacious argument: *paralogismos*
fallacy: *paralogismos*
fate: *heimarmenê*
father (v.): *gennan*
finite: *peperasmenos*
finite, be: *perainesthai*
fire: *pur*
fit (a.): *epitêdeios*
fitness: *epitêdeiotês*
flux: *rhusis*
follow (upon): *akolouthein, hepesthai*
for ever: *aei*
forecast: *katarkhê*
foreknow: *progignôskein*
foreknowledge: *prognôsis*
form: *eidos*
Form: *eidos, idea*
formative: *eidopoios*
frame: *sunistanai*
fraud: *skeuôrêma*
free of any relation: *askhetos*
free of need: *anendeês*
from everlasting: *ex aïdiou*
fulfilment: *teleiôsis*
furnish: *khorêgein*
future (a.): *mellôn*
future (n.): *to mellon*

generally: *haplôs*
generate: *gennan*
generated: *genêtos*
generation: *genesis*
give being: *ousioun*
go hand in hand with: *sumparathein*
go on to say: *epagein*
goal: *skopos*

god: *theos*
God: *theos*
going under the same name: *homônumos*
good (thing), (n.): *to agathon*
Good Itself: *autoagathon*
Good, the: *to agathon*
goodness: *agathotês*
grammar: *hê grammatikê*
grammar, of: *grammatikos*
grammarian: *ho grammatikos*
grammatical: *grammatikos*
grow (intrans.): *auxesthai*
growth: *auxêsis*

harmony: *harmonia*
have (its) essence: *ousiousthai*
have existence: *huphistasthai*
have prior existence: *proüphistasthai*
having the capacity to build: *oikodomikos*
heaven: *ouranos*
heavenly: *ouranios*
house: *oikia, oikos*
human: *anthrôpinos, anthrôpos*
hypothesis: *hupothesis*
hypothesise: *hupotithenai*

idea: *hupolêpsis*
identical: *aparallaktos*
ignorance: *agnoia*
ignorant, be: *agnoein*
illuminate: *katalampein, phôtizein*
illuminating, of: *phôtistikos*
illuminative: *phôtistikos*
illustration: *eikôn*
imagination: *phantasia*
imbue with form: *eidopoiein*
immaterial: *aülos*
immortal: *athanatos*
imperfect: *atelês*
imperishable: *aphthartos*
impervious to change: *apathês*
impiety: *asebeia*
impious: *asebês*
imply: *eisagein*
impossible: *adunatos*
impulse: *rhopê*
in (its) own right: *kath' hauto*
in a word: *haplôs*
in contact, be: *haptesthai*
in general: *holôs, haplôs*
in time: *khronikôs*
inadequacy: *astheneia*
inappropriate: *atopos*
incomplete: *atelês*
incorporeal: *asômatos*
indefinitely: *eis apeiron, ep' apeiron*
indicate: *noein, sêmainein*
indication: *apodeixis, dêlôsis*
indicative: *dêlôtikos*
individual (a.): *atomos*
individuals: *ta atoma*
inexperienced: *apeiros*
infer: *eisagein*
infinite: *apeiros*
infinitely powerful: *apeirodunamos*
inform: *eidopoiein*
ingenious: *sophos*
ingenuity: *deinotês*
inhere in: *enuparkhein*
initiative: *hormê*
inquire: *zêtein*
inquiry: *zêtêsis*
inseparable: *akhôristos*
instant (n.): *sêmeion*
instantaneous: *akhronos, athroos*
instantaneously: *athroos, athroôs*
instruct: *didaskein*
instruction: *didaskalia*
instrument: *organon*
intellectual: *noeros*
intelligible (n.): *to noêton*
intelligible (a.): *noêtos*
intelligible entity: *to noêton*
intend: *noein*
intention: *dianoia, nous, skopos*
interpret: *metalambanein*
interval: *diastasis*
introduce: *eisagein*
involuntary: *aboulêtos*
irrational: *alogos*
irreverent, be: *asebein*

keep pace with: *sumparathein*
kind: *eidos, genos, idea*
know: *ginôskein, noein*
knower: *epistêmôn*
knowledge: *epistêmê, gnôsis*

lack of movement: *akinêsia*
last (v.): *exarkein*
learn: *manthanein*
learned: *sophos*
learning: *mathêsis*

life: *bios, zôê*
life, to have: *zên*
light (n.): *phôs*
light up: *katalampein*
live (v.): *zên*
living creature: *zôion*
living-creature-itself: *autozôion*
local: *kata topon*
luminous: *phôtoeidês*

maintain (sc. preserve): *sunekhein*
make: *poiein*
make direct contact with: *prosballein*
make good: *agathunein*
man (sc. human being): *anthrôpos*
manifestly: *enargôs*
manner: *tropos*
material: *enulos*
mathematical science: *mathêma*
matter (n.): *hulê*
mean: *noein, sêmainein*
meaning: *dianoia*
measure (v.): *metrein*
measure (n.): *metron*
measure of: *metrêtikos*
mental sphere: *to noêtikon*
metaphor: *metaphora*
metaphorical: *metaphorikos*
method: *tropos*
middle (n.): *mesotês*
mind (n.): *nous, psukhê*
mobilisation: *prokheirisis*
mobilise: *prokheirizesthai*
mode: *tropos*
moon: *selênê*
moon, of the: *selêniakos*
moon's: *selêniakos*
mortal: *thnêtos*
move (intrans.): *kineisthai*
movement: *kinêsis*
movement, be in: *kineisthai*
mover: *to kinoun*

named for: *homônumos*
natural: *phusikos*
naturally: *phusikôs*
nature: *huparxis, phusis*
necessarily: *anankaiôs*
necessary: *anankaios*
necessity: *to anankaion*
negation: *apophasis*
negative (n.): *apophasis*
notion: *hupolêpsis, huponoia*
notionally: *kat' epinoian*
novice: *apeiros*
number (n.): *arithmos*

object of desire: *to orekton*
object of knowledge: *to gnôston*
object of thought: *to noêton*
observe: *horan, skopein*
obvious: *enargês*
occultation: *epiprosthêsis*
of any kind at all: *holôs*
of dialectic: *dialektikos*
of the mind: *noeros*
One, the: *to hen*
ontology: *philosophia ontôn*
opinion: *doxa*
opposed, be: *antikeisthai*
opposite: *to antikeimenon*
order (n.): *taxis*
organ: *organon*
origin: *arkhê*
outcome: *apotelesma*

parallel: *eikôn, hupodeigma*
parcel out along with: *sundiairein*
part (n.): *meros*
partake: *metekhein*
particular: *merikos*
passage: *khrêsis, lexis*
passing out of existence: *phthora*
pattern: *paradeigma*
pay reverence to: *eusebein*
perceptible: *aisthêtos*
perception: *antilêpsis*
perfect (a.): *teleios*
perfect (v.): *teleioun*
perfected: *teleios*
perfecting: *teleiôsis*
perfection: *to teleion, teleiotês*
perfective: *teleiôtikos*
period: *paratasis*
perish: *phtheiresthai*
perishable: *phthartos*
permanence: *diamonê*
philosopher: *philosophos, ho sophos*
philosophy: *philosophia*
physical: *organikos, phusikos*
pious: *eusebês*
place: *topos*
plant: *phuton*
plausibility: *pithanotês*
plenitude: *plêrôma*

point: *sêmeion*
portrait: *eikôn*
position: *dogma, doxa*
possession: *hexis*
possible, be: *endekhesthai*
potential: *dunamei, dunamis*
potentiality: *dunamei, dunamis*
potentially: *dunamei*
power: *dunamis*
precision: *akribeia*
predicate of: *katêgorein*
pre-exist: *proüparkhein*
premiss: *arkhê, protasis*
present (v.): *prokheirizesthai*
present, be: *huparkhein, pareinai*
present in, be: *enuparkhein*
preserve: *diasôizein, sôizein*
prevent: *kôluein*
principle (n.): *arkhê, logos*
privation: *sterêsis*
produce (v.): *poiein*
producer: *poiêtês, ho poiêsas, to poioun*
product: *apotelesma*
production: *paragôgê*
productive: *poiêtikos*
proof: *apodeixis, epikheirêma*
property: *pathos, dunamis*
prophet: *theologos*
prove: *apodeiknunai, kataskeuazein*
provide for: *pronoein*
providence: *pronoia*
provident: *pronoêtikos*
providential: *pronoêtikos*
pupil: *mathêtês*
put together: *sunistanai*
puzzle (n.): *aporia*
puzzle over: *aporein*

quality: *poiotês, to poion*
question (n.): *zêtêsis*
quote: *paratithesthai*

race (n.): *genos*
raise a puzzle: *aporein*
rational faculty: *to logikon*
ray: *aktis*
realisation: *entelekheia*
reason (n.): *aitia, logos*
reasoning: *epikheirêsis*
rebuttal: *antilogia*
receive: *lambanein*
receptive: *dektikos*
refer to: *noein, sêmainein*
reference: *sêmasia*
refutation: *elenkhos, lusis*
refute: *elenkhein, luein*
region: *topos*
relation: *skhesis*
relatives: *ta pros ti*
remain: *hupomenein, menein*
remove: *aphairein*
represent: *kataskeuazein*
reproduce: *gennan*
responsible for: *aitios*
rest (n.): *êremia*
reveal: *deiknunai*
reversal: *metathesis*
revolution: *periphora*
ridiculous: *geloios*
rigour: *akribeia*

sacrilegious: *asebês*
say of: *katêgorein*
scientific: *phusikos*
scripture: *logion*
secondary origin, of: *husterogenês*
section: *logos*
see: *horan, sunaisthanesthai*
seeing: *opsis*
seek: *zêtein*
seem: *phainesthai*
self-subsistent: *authupostatos*
self-sufficient: *autarkês*
seminal: *spermatikos*
sensation: *aisthêsis*
sense: *aisthêsis*
sense: *nous*
sense (sc. meaning): *to sêmainomenon, tropos*
separate (a.): *kekhôrismenos, khôristos*
separate (v.): *khôrizein*
shadow (n.): *skia*
shadow, throw into: *skiazein*
share (in): *koinônein, metekhein*
show: *apodeiknunai, deiknunai*
sight (n.): *opsis*
signification: *sêmasia*
signify: *sêmainein*
simple: *haplous*
simply: *haplôs*
skill: *epistêmê*
skilled in building: *oikodomikos*
sometimes: *pote*
sophist: *sophistês*

sophistical: *sophistikos*
soul: *psukhê*
soul, of the: *psukhikos*
source: *arkhê*
source of illumination: *to phôtistikon, to phôtizon*
species: *eidos*
specify: *prosdiorizesthai*
sphere: *sphaira*
spring: *pêgê*
stability: *stasis*
state (v.): *apophainesthai*
state of affairs: *katastasis*
statement: *lexis, logos*
statesman: *ho politikos*
stationary: *ametabatos*
stop: *pauesthai*
stuff (n.): *hulê*
subject (n.): *to hupokeimenon*
substance: *ousia*
substratum: *to hupokeimenon*
subtlety: *deinotês*
successful: *eudaimôn*
successor: *diadokhos*
suffice: *exarkein*
sun: *hêlios*
sun's: *hêliakos*
superior: *kreittôn*
supply (v.): *khorêgein*
suppose: *hupotithenai*
supramundane: *huperkosmios*
survive: *sôizesthai*
syllogism: *sullogismos*

take: *lambanein*
teach: *didaskein*
teacher: *didaskalos*
teaching: *didaskalia*
temporal: *hupo khronon, khronikos*
temporally: *khronikôs*
tense (n.): *khronos*
theory: *theôrêma*
thing: *pragma*
think (about): *ennoein, noein*
thought: *dianoia, ennoia, epinoia, noêma, noêsis*
through an act of creation: *dêmiourgikôs*
time (n.): *khronos*
time, at a (at some): *pote*
time, of: *khronikos*
timelessly: *akhronôs*
to quote his exact words: *epi lexeôs*
tool: *organon*
touch (v.): *haptesthai*
transformative: *alloiôtikos*
transition: *metabasis*
transparent: *diaphanês*
traversable: *diexitêtos*
treatise: *logos, pragmateia*
trickiness: *deinotês*
true: *alêthês*
true at the same time, be: *sunalêtheuein*
true, be: *alêtheuein*
truth: *alêtheia*
turn: *periagein*

unacquainted: *apeiros*
unaware, be: *agnoein*
unchanging: *ametablêtos*
undergo: *hupomenein*
undergo alteration: *alloiousthai*
understand: *noein*
understanding: *nous*
undertaking: *hormê*
undivided: *adiairetos*
undo: *luein*
unerring: *aparalogistos*
ungenerated: *agenêtos*
uniform: *homoeidês, homoiomerês*
unique: *monogenês*
unit: *monas*
unitary: *heniaios*
unity: *to hen*
universal (a. or n.): *holikos, katholou*
universally: *katholou, en tôi katholou*
universe: *to pan*
unmoving: *akinêtos*
untraversable: *adiexitêtos*
usage: *khrêsis*
use (n.): *khrêsis*
use of: *katêgorein*

various: *diaphoros*
verbal: *logikos*
vice: *kakia*
view (n.): *doxa*
villainous way: *kakourgia*
visible, be: *phainesthai*
vision: *opsis*

water: *hudôr*
way: *tropos*

weakness: *astheneia, to asthenes*
well-being: *eudaimonia*
whole: *holos*
will (n.): *boulêsis*
will (v.): *ethelein*
will, act of: *boulêsis*
wish (v.): *boulesthai, ethelein*
with infinite power: *apeirodunamôs*
without any intermediary: *amesôs*
without beginning: *anarkhos*
without end: *ateleutêtos*
without mental imagery: *aphantastôs*
without origin: *anarkhos*
without parts: *amerês*
without qualification: *adioristôs*
without relation: *askhetôs*
word(s): *lexis, onoma, prosrêma*
work (sc. book): *logos, pragmateia*
work (sc. creation): *poiêma*
world (a.): *kosmikos*
world: *kosmos*

Greek-English Index

This index lists a selection of more important words from the Greek text together with my translations of them. I have not attempted to distinguish between Philoponus' own words and those of Proclus and other authors he quotes. The rubric 'other tr[anslation(s)]' covers cases where a word has been translated in such a way that there is no one-to-one correspondence between the Greek and the English. The page and line references are to Rabe's Greek text and the occurrence of 'etc.' at the end of a listing of such references means that it is incomplete.

Subject Index

For the most part I index topics rather than attempt to summarise arguments, but I have included extremely brief summaries of the arguments of Proclus under the entry for Proclus. (Summaries of the sections of Philoponus' refutations of these, perhaps by Philoponus himself, appear in the translation). I have also included all proper names that occur in the translation with a full list of occurrences or, in a few cases, an indication of the number of times they occur. References are to the page and line numbers of Rabe's Greek text, which are printed in the margins of the translation, those to Proclus' arguments, or to substantial quotations from other works of his, being in bold. An exception to this are references to Proclus' first argument, which take the form '**Arg.1**'.

Index of Passages Cited in the Notes

Numbers in bold type refer to the notes, those preceded by 'i' to the notes to the introduction.